CUTTING-EDGE BIOETHICS

A Horizons in Bioethics Series Book from

THE CENTER FOR
BI◉ETHICS
AND HUMAN DIGNITY

The Horizons in Bioethics Series brings together an array of insightful writers to address important bioethical issues from a forward-looking Christian perspective. The introductory volume, *Bioethics and the Future of Medicine,* covers a broad range of topics and foundational matters. Subsequent volumes focus on a particular set of issues, beginning with the end-of-life theme of *Dignity and Dying* and continuing with the genetics focus of *Genetic Ethics,* the economic and patient-caregiver emphases of *The Changing Face of Health Care,* and the reproductive analyses in *The Reproductive Revolution.* The volume on *BioEngagement* develops strategies for tacking such challenging issues.

The series is a project of The Center for Bioethics and Human Dignity, an international center located just north of Chicago, Illinois, in the United States of America. The Center endeavors to bring Christian perspectives to bear on today's many pressing bioethical challenges. It pursues this task by developing two book series, hundreds of audios and videos, numerous conferences in different parts of the world, and a variety of other printed and computer-based resources. Through its donor-member support program, the Center networks and provides resources for people interested in bioethical matters all over the world. Supporters receive the Center's international journal, *Ethics and Medicine,* the Center's newsletter, *Dignity,* the Center's Update Letters, an Internet News Service, and discounts on a wide array of bioethics resources.

For more information on membership in the Center or its various resources, including present or future books in the Horizons in Bioethics Series, contact the Center at:

The Center for Bioethics and Human Dignity
2065 Half Day Road
Bannockburn, IL 60015 USA
Phone: (847) 317-8180
Fax: (847) 317-8101

Information and ordering are also available through the Center's World Wide Web site on the Internet: *www.cbhd.org.*

THE CENTER FOR
BI◉ETHICS
AND HUMAN DIGNITY

CUTTING-EDGE BIOETHICS

*A Christian Exploration
of Technologies and Trends*

Edited by

JOHN F. KILNER,

C. CHRISTOPHER HOOK,

&

DIANN B. UUSTAL

WILLIAM B. EERDMANS PUBLISHING COMPANY
GRAND RAPIDS, MICHIGAN / CAMBRIDGE, U.K.

Published 2002 by Wm. B. Eerdmans Publishing Co.
255 Jefferson Ave. S.E., Grand Rapids, Michigan 49503 /
P.O. Box 163, Cambridge CB3 9PU U.K.

Printed in the United States of America

07 06 05 04 03 02   7 6 5 4 3 2 1

ISBN 0-8028-4959-8

www.eerdmans.com

Contents

II. GROWING CULTURAL CHALLENGES

III. THE CHANGING FACE OF HEALTH CARE

IV. PROACTIVE PERSPECTIVES

Contributors

Mary B. Adam, M.D., Clinical Lecturer, Department of Pediatrics, University of Arizona College of Medicine, Tucson, AZ, USA

Helen Alvare, J.D., Media Consultant; Professor, George Washington University Law School, Washington, DC, USA

Linda K. Bevington, M.A., Director of Research, The Center for Bioethics and Human Dignity, Bannockburn, IL, USA

Francis S. Collins, M.D., Ph.D., Director, National Human Genome Research Institute, National Institutes of Health, Washington, DC, USA

E. David Cook, Ph.D., D.Litt., Director, Whitefield Institute for Research, Oxford University, Oxford, England

Robert K. Garcia, M.A., Instructor, Torrey Honors Institute, Biola University, La Mirada, CA, USA

Francis Cardinal George, O.M.I., Ph.D., Archbishop, Roman Catholic Church, Chicago, IL, USA

C. Christopher Hook, M.D., Hematologist/Ethicist, Mayo Clinic, Rochester, MN, USA

John F. Kilner, Ph.D., President, The Center for Bioethics and Human Dignity, Bannockburn, IL, USA

CONTRIBUTORS

Nancy L. Jones, Ph.D., Assistant Professor of Pathology, Wake Forest University School of Medicine, Winston-Salem, NC, USA

C. Ben Mitchell, Ph.D., Associate Professor of Bioethics & Contemporary Culture, Trinity International University, Deerfield, IL, USA

Dónal P. O'Mathúna, Ph.D., Professor of Bioethics and Chemistry, Mount Carmel College of Nursing, Columbus, OH, USA

John Patrick, M.D., Professor, University of Ottawa, Ottawa, Canada

Scott B. Rae, Ph.D., Professor of Biblical Studies and Christian Ethics, Talbot School of Theology, Biola University, La Mirada, CA, USA

Daryl Sas, Ph.D., Associate Professor of Biology, Geneva College, Beaver Falls, PA, USA

Richard A. Swenson, M.D., Director, Future Health Study Center, Menomonie, WI, USA

Diann B. Uustal, R.N., Ed.D., President, Educational Resources in Health-Care, Jamestown, RI, USA

Introduction

This moment in history is a crucial point for the human race. New technologies promise vast improvements in health care and for the first time genuinely present the possibility of overcoming major disabilities such as blindness and paralysis. But at the same time these technologies may prove devastating: promoting loss or erosion of personal identity, tightening the new shackles of an ever more powerful technological tyranny, or even contributing to the destruction of our species.

Genetics, cybernetics, and nanotechnology, for example, each promises to reverse or eliminate many disease states and physical limitations. Yet each may also be used for augmentation purposes to pursue the engineering of "better" human beings. If augmentation is permitted and uncritically embraced by medical and other communities, discrimination against the unenhanced will quickly follow. In a culture already deeply fragmented, such technologies — if not appropriately restricted — could lead to the further breakdown of the social matrix. At this point, there is no clear set of definitions demarcating the line between disease and trait — between therapy and augmentation. Meanwhile, tools such as nanotechnology and genetic engineering which can manipulate matter at the most minuscule level may be marvelous tools to treat cancer, coronary artery disease, and certain infectious diseases. Yet any tool that can operate at this level could also be turned into a new plague to destroy rather than heal.

Most frightening about our day, however, is not the development of new technologies. Rather, it is the fact that technologies with this much power are arising at a time when humanity may not be capable of developing them responsibly. We find ourselves in an environment increasingly relativistic, morally adrift, and hostile to God. A true understanding of human nature and our responsibilities to our Creator and each other has been replaced with a materialistic nihilism giving birth to a school of thought called transhumanism. Transhumanism, or post-humanism as it is sometimes called, has an open contempt for God's created order, believes that humanity is something to be overcome and replaced by our technologies, and advocates that a new species or several new species of techno sapiens should be created to fulfill our "evolutionary destiny." While most individuals today are not transhumanists, the uncritical adoption of new technologies will enlist more and more individuals as unwitting participants in the fulfillment of transhumanist goals.

This brings us to the importance of the present volume. If humanity is to survive in the future, we must reverse the trends of the past. We cannot continue to be surprised by new technologies — forced to scramble to perform the ethical analysis and implement means of control. We must prospectively engage technologies that are surely coming, doing the ethical analysis now and proposing and implementing the safeguards before the technology is unleashed. This book is therefore intended as a wake-up call, an invitation to look to the near future and see what is coming soon so that we can carry out the critical task of analysis and engagement. At the same time as it helps prepare us for tomorrow, this book is also designed to strengthen our engagement of cutting-edge challenges in health care today — issues that have warranted more widespread attention long before now. Another major goal of this book is to demonstrate the critical importance of God's perspective in the evaluative process. Only as we respect God's design and plans for humanity can there be any long-term hope of protecting and preserving human significance and dignity — in fact the very species homo sapiens, the bearer of the image of God.

The book opens with a section on emerging technologies, in order to establish at the outset a concrete sense of what is possible today and likely in the future. Francis Collins, head of the Human Genome Project to map the human genetic code, begins the analysis with an examination of progress in genetics and its ethical implications. Oxford University's David Cook follows with an exploration of using animal body parts in humans, so-called xenotransplantation. Scientists Nancy Jones and Linda Bevington then consider related issues at the more minute level of transgenics: mixing the genetic material from various species, including humans. This opening section also looks farther into the future, beginning with researcher Robert Garcia's account of the directions that artificial intelligence efforts are taking. To what extent will human beings ultimately be necessary in the world of tomorrow?

Mayo Clinic's Christopher Hook joins Garcia's consideration of that question by looking at how mechanical technologies will increasingly replace bodily functions and activities, courtesy of the rapidly developing fields of cybernetics and nanotechnology. What emerges from this section is a clear sense of the awesome significance of what is unfolding in the realm of biotechnology — and the importance of addressing the ethical challenges today.

In order to meet such challenges, however, a clearer understanding of critical contextual issues such as culture, technology, and ethics is essential. The book's second section is devoted to furthering such understanding. The section begins with medical ethicist John Patrick's investigation of the multiculturalism of today's world. Widely traveled, and based outside the United States, Patrick brings an international perspective to some of the cultural dynamics involved in translating scientific developments into medicine. Theological ethicist Daryl Sas then homes in on the issue of technology itself, formulating insights that can help us grapple with the particular technologies of our day, such as stem cell research. Francis Cardinal George concludes this section with a consideration of the ethical vision necessary if we are to be able to resist destructive temptations and forces and channel scientific and medical developments in ways that will truly be a blessing to the human race.

While major developments are under way in science — with great implications for health care — there are also hugely significant cutting-edge developments today in health care itself that warrant special attention. The book's third section tackles such developments first through business ethicist Scott Rae's analysis of economic and business forces that are shaping the direction of health care in important ways. Next, nursing specialist Dónal O'Mathúna evaluates emerging trends in spirituality and alternative medicine. Physician Mary Adam then singles out for careful consideration an ethically-charged area of medicine particularly in need of greater insight and initiative: the prevention of sexually transmitted diseases such as AIDS. Health educator Diann Uustal concludes the section with a critical assessment of the state of "care" in health care today — and what needs to be put in place for tomorrow.

Armed with an understanding of emerging technologies, growing cultural challenges, and the changing face of health care, we cannot help but wonder what we are to do with this understanding. The fourth and final section of the book addresses this question in three ways. First, futurist Richard Swenson offers an overview of the broad range of developments in science and medicine and considers how to cope with such a plethora of change. Legal and media specialist Helen Alvare then speaks to the media and public policy challenges and opportunities created by scientific and medical developments. The book concludes with theological ethicist Ben Mitchell's clarification of what is at stake ethically in all of this, and the kind of proactive engagement that is necessary if science and medicine are to make good on their promises at a price that humanity can afford to pay.

Needless to say, any work such as this is a team effort in ways far from obvious. Those acknowledged here are representative of a far greater array of people who have had some hand in carrying out this project. We are deeply grateful to Jay Hollman, without whose support and vision this project would not have been possible. Jon Pott and the Eerdmans staff were of tremendous help in shepherding this work through the editorial process, as was Laura Hepker at The Center for Bioethics and Human Dignity. Reviewers too numerous to mention worked with the book's outstanding team of authors to produce the final text printed here. And we would be remiss if we did not also give ultimate thanks to God — Creator, Healer, and Comforter — who offers profound wisdom for tackling the awesome bioethical challenges facing the human race. Though we have undoubtedly grasped here but a small portion of this wisdom — and that imperfectly — it is our hope that this book and the ongoing work of the Center will encourage an aggressive search for the wisdom that humanity needs for a healthy future.

John F. Kilner, Ph.D.
C. Christopher Hook, M.D.
Diann B. Uustal, R.N., Ed.D.

PART I

EMERGING TECHNOLOGIES

Human Genetics

FRANCIS S. COLLINS

The time is ripe for a serious conversation about the pathway that genetic science is leading us down.[1] It is a pathway that I as a physician and as a Christian have a great deal of hope about, because of its promise for alleviating human suffering, which is surely one of our strongest mandates. However, it is also a pathway which poses certain troubling risks. Such risks are real possibilities that we must attempt to address effectively. When we have done so, we can move this exciting field forward in a way that maximizes the benefits and minimizes the risks.

While many of the issues that biomedical research raises are vexing, they also have solutions. Some of the scenarios that sound truly scary are actually not scientifically very likely. However, there are some that are both realistic and quite troubling. Still, the bottom line is that the promise of this research is so great for medicine that the most unethical thing we can do is to slow it down. Of course, this creates a certain responsibility for all of us. We need to be sure that as we move research forward at the maximum speed for the sake of its benefits, we also make sure that we are attending to the social, ethical, and legal issues that are part of this enterprise.

The Human Genome Project

While people have compared the Human Genome Project to many things, one of the most common analogies is to the Lewis and Clark expedition. This

1. This chapter is an edited transcript of an oral presentation meant to encourage such conversation.

time, however, we are engaged in the exploration of ourselves, not the Northwest Territory. It is a historic undertaking: we only do this once in our history. In studying the complete set of genes that we carry, we are trying to learn the "parts list" for human biology. We are exploring our instruction book. Historically and philosophically, this is profound. Our human biological instruction book allows us to carry out all of our biological activity from the time that we are single-celled embryos until the end of our lives. It is an exciting notion that we now have this instruction book in front of us and are able to read it, even if we don't understand it very well.

Nevertheless, we must also address the issues raised by this project that could be potentially harmful. One Time/CNN poll reports that a significant number of people are really troubled about the Human Genome Project. Among those surveyed, 46% said that the project would generally be harmful. Interestingly, however, 61% said they would like genetic information about future disease risk. But such information will be accessible only if the Human Genome Project is carried out successfully, so there is a certain discordance here. The poll, however, did not ask whether people actually know what the Human Genome Project is. I have done my own polls on that question for four or five years, and I am sorry to report that until about a month ago, only one out of three people had any real idea. So we need to understand why the project has such negative connotations for the layperson.

Part of the problem is that we have not been very successful at explaining the actual science and purpose of this project in a fashion that makes sense to the public. It seems impenetrable, because it's covered over with a lot of jargon. We scientists are terrible about that, and the project hasn't been presented or portrayed in an easily understandable manner. The poll mentioned above certainly shows that there is a large amount of work to be done in that direction.

Genetics is the best tool we have yet found for unraveling the mysteries of a disease. Many of the diseases that we wish we understood well enough to eliminate still elude our best efforts to do so. Most diseases come about as consequences of interactions between hereditary factors and the environment. The course of a disease may be altered by medical treatment or by the nutrients ingested, so one shouldn't think of any disease as completely genetic. At the same time, aside from trauma, few conditions are completely environmental in origin. With diabetes, there is roughly equal influence between environmental factors (many of which are unidentified) and genetic factors. Most of the common illnesses that are afflicting so many of us — cancer, heart disease and mental illness — are similar to diabetes.

One would think that at least an infectious disease like AIDS, which after all is caused by a virus, ought to be called an environmental disease. However, it is not as easy as that, because host factors that are encoded by our genes will play a significant role in what happens if we are exposed to this virus. For example, in a room of several hundred people, there are probably several people who are genetically immune to getting AIDS. They could be exposed to the virus over and over again, but they would never acquire the disease because they genetically lack a cofactor that is necessary for the virus to get into the cell. If we could figure out how to transmit that particular resistance into other people, we would have a wonderful vaccine.

Still, in the end, it may seem that we should be devoting as much effort as possible to trying to understand the environmental causes of disease as we are to analyzing the genetic causes. Many are doing so. However, environmental causes are significantly harder to study. Many of us have actually come to the conclusion that the best way to understand the environmental effects is to understand genetics, because environmental risks are not shared by all of us. I may be put very much at risk by some exposure that may not jeopardize you at all. Until we begin to understand our varying reactions, we may not really understand what environmental risks are about either. It's the interaction that we need to understand, and genetics gives us powerful tools to that end. Therefore, principle number one is that all diseases except trauma have a genetic component, and approaching disease by this particular entry way is going to explain many mysteries.

Principle number two is that we all have genetic defects. If we think we're the perfect genetic specimens, we are in for a great disappointment. We are all carrying something like forty to fifty faulty genes around in our systems. Most of those will not become apparent because we don't have the mixture of genes and environment to actually cause a disease to appear. However, we are all flawed. This is the biological equivalent of original sin: we have all fallen short. So this means that genetics is not just the study of rare diseases that happen in other people's families. This is about all of us.

To understand the hereditary factors, we must understand the wonderful molecule called DNA, the double helical structure that Watson and Crick figured out some forty-seven years ago. It is a very elegant system for encoding information. What a privilege it is for a physician-scientist to do research that uncovers something about our creation and gives us a glimpse into the elegant way God thinks! DNA is certainly a remarkable way of coding information in a very efficient and digital fashion, allowing us to carry around an enormous amount of information in a very modest space.

In each individual's genetic code, there are three billion base pairs, where each base can be either adenine (A), cytosine (C), guanine (G) or thymine (T). A always pairs with T and G always pairs with C. The four possible choices for each one of these three billion positions yield a huge amount of potential coding capacity. In light of all the things that we have to do as human beings, three billion base pairs may not seem like quite enough, but it must be because it works. Astonishingly, much of our DNA does not have an obvious function. An estimated seventy percent or more of it does not contain genes and just seems to be along for the ride. However, we will never know which part of the genome is the functioning part without studying all of it, and this has justified the broad, all-encompassing approach to the Human Genome Project.

The project began approximately ten years ago with the goal of determining the genetic map of human beings as well as of a number of other organisms. Studying other organisms provided the opportunity for cross comparison in order to learn something about how genes work and how biology is connected to them. The project didn't really get into sequencing all those three billion letters until about four years ago. Up until then, we built maps of the human chromosomes, basically putting "mile markers" along the chromosomes. We began to sequence the human genome in about 1996, and it was tough going at first. There was a low point about three years ago when the goal looked like it was going to be extremely difficult to achieve, but then things started to happen. The technology got better, and scientists built better automated systems. The centers that were competing with each other agreed to work together, and I found myself as the project manager of a group of sixteen centers in six countries that were all ready to collaborate to make sure the whole project got done.

So the target date — to complete the sequence in 2005 — began to move up. It really began to move during the spring of 1999, when a private company announced its intent to sequence the human genome and claimed that it could do it faster, better, and cheaper than others. This competitive element was good, because it did stir the pot a bit and got people's blood flowing. However, the real reason for the acceleration of this timetable was largely the advance in technology achieved through the public project, a fact that set this private enterprise in motion. The public and private approaches were quite different and generally complementary, and both achieved significant progress. As a result, on June 26, 2000, in a perhaps overdone but rather inspiring moment involving the White House and 10 Downing Street, both the public and the private efforts made an announcement that significant milestones had been reached in these enterprises: ninety percent of the sequence was now in hand.

The public Human Genome Project has since published the human sequences on the Internet where anyone can go and see them. It has been our principle from the beginning to give this information away as soon as it becomes available. The sequence of the human genome is something that should not be kept private. It should not be tied up as the intellectual property of any particular university or company. It ought to be accessible so that anybody with a good idea can study it and try to understand how it works. It is our common inheritance, and it ought to be available to anyone in the world.

The Human Genome Project made much progress in little more than a year. Many of the chromosomes hadn't been touched as of March 1999, and we had done only about fifteen percent of the sequencing work. But fifteen months later, in June 2000, ninety percent of the Human Genome Project had been made available in the public data bank in either finished or draft form, accessible to anybody who wanted to look at it.

The draft form still has gaps that need to be cleaned up, and it has ambiguous places where we are not sure if something is, for example, a T or an A. That needs to be corrected, because, over the course of time, this ought to be a sequence that is done absolutely right. However, most of the questions that scientists want to ask about the DNA sequence are fairly well answered with the draft sequence, which is why we rushed to produce it as quickly as possible.

It will take another couple of years to have all of the sequences in completely finished forms. Chromosomes 21 and 22 were completely sequenced as of June 2000, but the rest of the project needs to be completed as well. Doing so will take the continued public collaboration of sixteen centers until the job is done, hopefully by the year 2003.

Next Steps

There is, of course, a question about the proper roles of public and private efforts in mapping the human genome. This is not a trivial issue to work through. The benefits of genomic research cannot be expected to reach the public in the form of advances in medicine without vigorous involvement by the private sector, including biotechnology and pharmaceutical companies. Almost every pharmaceutical company today has a genomic division, because that will be the source of the next generation of drugs. Biotechnology companies have sprung up in great abundance and are doing wonderful things to speed up the process of translating the basic information about our DNA sequence into advances in diagnostics and therapeutics.

This partnership between the public and private spheres needs to be encouraged and nurtured. At the same time, one has to be careful about the ways in which access to something as important as the genome sequence, or to various bits and pieces of it, becomes constrained by intellectual property laws. There is a vigorous debate about the proper role of patents. The real question is what is good for the public. If a patent will ultimately be good for the public, then it is commendable. If it gets in the way of public well-being by creating a disincentive for other groups to work on something because it has already been claimed, then it is inappropriate. Most observers agree that there is a role in certain circumstances for patents on genes because they provide an incentive for research that otherwise will not exist. We need to be very cautious about granting patents on genes whose function we don't understand.

Mapping the genetic code is not, of course, the end of the story. Reading out the sequence and having it in front of us is really only the end of the beginning. The interesting and challenging part is figuring out what it all means. These three billion letters seem to make up somewhere in the neighborhood of thirty thousand genes, each gene being a packet of information that conveys a certain instruction. The fact that we're not even sure what the actual number of genes is, even though we have most of the sequence in front of us, shows just how hard it is to stare at page after page of A, C, G and T and figure out what it's telling us. We have a lot of work to do to understand this language. If we printed out the whole sequence, and piled the pages on top of each other, these three billion letters would be as high as the Washington monument. So we have a large book written in a language that we don't yet understand very well.

One of the ways in which we are making headway is by looking at human variation. Human beings are quite similar in terms of genetics. Comparing any two people's DNA sequences letter by letter, we may encounter a difference about every one thousand letters where one person may have a C and the other may have a T. The other 999 letters would be the same. The two people's DNA would be 99.9% identical, regardless of ethnicity or population background. Nevertheless, that 0.1% still means that there are many variations between these people's genomes. If there are three billion base pairs, there might be three million differences. If large numbers of people are in view, there are going to be even more differences. Roughly ten million common sites of variation would be found. Most of those ten million differences between us probably fall in parts of the genome that aren't doing very much, so the differences don't have consequences. The numbers of common variants that actually affect function may be as small as a couple of hundred thousand, many of which have rather mild effects. On the other hand, a few of those will fall into those vulnerable parts of the genome that can put people at risk for diabetes or Alzheimer's disease. Those are the ones that we are most interested in uncovering.

The cutting edge of genetic research is to look at all of the human genes to see whether the spelling of each is correlated with disease. An example of how this works in two genes, gene 1 and gene 2, is as follows. Gene 1 has two different spellings — one with a G in a particular position, and one with a C in that same position. People with diabetes have no greater likelihood of having either a G or C than unaffected people. There doesn't seem to be a correlation here; the frequency is the same, so the variation in gene 1 doesn't seem to be playing a very large role in the risk of diabetes.

Gene 2, on the other hand, is much more interesting. It has two spellings — a T or an A in a particular position, and the T appears much more frequently in the affected and the A in the unaffected. That would make scientists wonder if the T spelling of gene 2 is one of the predisposing factors for diabetes. They would need to do some mathematics and analyze the statistics to be sure they are not just fooling themselves. We've never had the ability to do this before in a systematic way. In the past, we have had to guess, and our guesses were almost always wrong because we didn't understand disease well enough to know what a good candidate gene would look like. However, now we can just look at all of the genes, because the genome is a bounded set of information; it's not going to get any bigger. The answers are in there, and this kind of analysis will reveal genetic components for most of the common disorders that affect us. Within the next five years or so, we will know what the hereditary factors are in most common diseases.

Why are we doing all this? The New Testament book of Matthew serves as a reminder of how much time Christ spent healing people in his very short time on this earth: "Jesus went through all the towns and villages, teaching in their synagogues, preaching the good news of the kingdom and healing every disease and sickness" (Matthew 9:35). Perhaps because they are called to be Christlike, Christians feel a particular responsibility for reaching out and healing the sick. That is one of the reasons why studying this aspect of our biology and trying to apply it medically is not merely a good idea, but a moral necessity. It's an ethical requirement of us. If we have the ability to heal, if genetic research holds out hope and promise, if it can benefit individuals for whom we have nothing to offer otherwise and not harm others in the process, then we have to do it. But we do have to shoulder the responsibility of making sure it is used for good purposes and not for unethical ones.

Coming Medical Advancements

How are these discoveries about the genome going to affect medical care? One way we can help people is through identifying (cloning) disease genes. We would clone a gene by purifying a piece of DNA away from the rest of the genome because we know that it is the piece responsible for a particular medical problem. This is not cloning people. We don't do that in the Human Genome Project. Cloned genes will help us develop diagnostic tools to detect whether a given person has a genetic predisposition for a particular disease. We have such tests now for breast cancer, colon cancer, Alzheimer's disease, and a fairly short list of other relatively common diseases. That list is going to grow quickly. However, is this necessarily a good thing, particularly if there is as yet no intervention to treat a problem?

How many of us really want to know what's lurking in the future for us? If we have forty or fifty gene defects — all of which are of the "there is nothing that we can do about it but we thought you might want to know" type — such knowledge is rather unappealing. On the other hand, if an intervention is already available for a certain disease, then such knowledge may be very desirable, because it would allow us to do something to decrease our chances of getting a terrible disease. We are already getting close to this level of knowledge for some conditions such as colon cancer. Diagnostics are most useful if they are associated with a preventive medical strategy that people can undertake to reduce their risk. This is preventive medicine, but it is individualized instead of "one size fits all."

Pharmacogenomics is going to be another beneficial development. This is the effort to identify why it is that people don't always respond in the same way to the same drug given for the same disease. For instance, a drug that is the drug of choice for a certain disease may not always have an effect on the disease. Or worse yet, it may give the patient a side effect that is unexpected and makes the illness worse. Why does this happen? The most probable answer put forth so far is that these variations are caused by the inherent differences coded into our DNA. If we could make predictions about how these differences affect different treatments, we could choose drugs more effectively.

What people are most excited about, however, is our potential ability not only to make predictions but also to discover actual cures, so that if people fall through the prevention safety net, something will still be there to catch them. Gene therapy gets much attention in this regard, because it is radically different from other more traditional approaches to medicine. In its simplest form, gene therapy involves transferring a normal copy of the gene that is not working to the tissue where its action is needed. For cystic fibrosis,

10

for example, transferring a normal copy of the gene to the lungs of an affected young person might be expected to provide considerable benefit, if the gene could be transferred efficiently to a high proportion of the cells in the lung, and if it was properly regulated. Unfortunately, this approach has not yet shown much in the way of concrete clinical benefits and is not free of risks. The death of Jesse Gelsinger during a 1999 gene therapy trial reverberated through the scientific community and the public at large. It was the end of our innocence about gene therapy.

We have a long way to go to develop gene therapy fully, but it should eventually produce effective treatment for certain diseases. Hemophilia, for example, may be near the top of that list. But gene therapy will most likely not be the solution for a host of other conditions. In some cases, it will be extremely difficult to deliver the gene — for instance, to the brain. We should definitely be pursuing gene therapy at the basic science level with maximum intensity because it does have much promise. However, we also have to be realistic in recognizing that many of the promises of gene therapy have leapt ahead of actual reality.

Discovering the genetic basis of disease provides another way to develop new therapies. For many diseases, this information will help us understand the basic biological problem that is at the root of a particular condition. We do this by going to the gene level and understanding the molecules the gene produces. This understanding then gives us an idea about the fundamental problem. That allows the opportunity to design a drug that will address the problem.

That kind of gene-based drug design is already beginning to happen in a few specific instances, although most applications still lie a few years ahead. One particularly intriguing example, which seems to be working rather dramatically, has to do with chronic myelogenous leukemia (CML), one of the common types of adult leukemia. In most people with this disease, there is a rearrangement between chromosome 9 and chromosome 22, which break and rejoin in a particular way, forming what is called the Philadelphia chromosome because it was first identified by an investigator in Philadelphia. The arrangement brings together a part of chromosome 9 and a part of chromosome 22 that shouldn't be next to each other, and creates a fusion between two different genes, one of them called BCR and the other called ABL. This fusion gene produces a product that can take normal white blood cells and cause them to become malignant.

What exactly is it about this fusion gene that produces such a dangerous result? If DNA makes RNA, which then makes protein, what is the protein product of this gene doing? It turns out that the resulting protein is a tyrosine kinase, an enzyme that catalyzes a particular biochemical reaction. Just in the last couple of years, researchers at Novartis have understood the protein product enough to recognize that it transfers a phosphate from the molecule called ATP to some substrate, and that starts a cascade, which converts a normal cell to a malignant cell. In response to this knowledge, they designed a drug (originally called STI571) that fits precisely into a pocket of the BCR/ABL protein so that the ATP can't get in there. It effectively blocks the action of the enzyme. In other words, it is a designer drug based upon a molecular understanding of the problem in this type of leukemia. Of the first thirty-two patients who received this drug in a phase one clinical trial (all patients who had failed all other forms of therapy and were not expected to live for more than a few weeks), all but one went into a complete remission. That result appears very rarely in new drug trials, and it has many people electrified. Undoubtedly, this will not turn out to be the solution for every leukemia patient, but the prospects are extremely exciting. This is the kind of pathway that we hope can be traveled over and over again for disease after disease, because problems can be understood at this level of specificity.

A Likely Scenario

Now imagine another scenario. The year is 2010. Our patient, John, is twenty-one, and in a routine screening physical, his employer has discovered that he has a high blood cholesterol level. He also smokes a pack of cigarettes a day and has done so for the last five years. John's father died of a heart attack at the age of forty-eight. His mother is alive and well at age fifty-seven, but a number of other family members have heart disease.

As much has been learned by this point, his physician offers him tests to help figure out why John's cholesterol is high. Other tests can determine his risk of other problems, including things that might not clearly run in his family but would matter to him. Is he interested? Let's say John doesn't want to hear about anything that he can't do something about, but he is interested in identifying problems that he might be able to prevent. So John is given an educational video to take home in order to figure out which of these tests he wants to have. He comes back with his requests, including one that the physician figure out why his cholesterol is up, because he is somewhat worried.

John is at a very important teachable moment. He has not previously thought much about his health, but now for the first time he is concerned. This could be a great moment to set him on a pathway toward a healthier lifestyle. The results of the DNA tests that are done on John's blood will be complex. These are not tests that will say, "you are going to get this disease," or "you will never get that one." They will be couched in statistical terms. John will be told that his risk is elevated, or that it is relatively low, and he will need to understand relative risks and absolute risks.

His results may be a mix of good news and bad news. Genetic tests don't always produce bad news. Our DNA codes contain much information that is really positive. For example, some of us are actually quite resistant to certain diseases because of the spellings we carry around. It turns out that John has inherited some good gene spellings for preventing prostate cancer and Alzheimer's disease and so his relative risks for those are down at 0.4 and 0.3, (with 1.0 being the "norm"). His lifetime absolute risks are not zero — they are 7% and 10% — because these are such common disorders. So, though this is not quite as reassuring as John had hoped, it is still good news relative to the average risk.

The next stage is the one that may require more action. One test has revealed why John's cholesterol is elevated, and he is indeed at risk of coronary artery disease. He has spellings of the genes called Apo B and CETP that elevate his risk of heart disease 2.5-fold over the average person, giving him a 70% chance of symptomatic heart disease by age seventy. This is a pretty significant issue and is undoubtedly related to his elevated cholesterol and to his father's heart disease. More surprisingly for him, he has a fourfold increased risk of colon cancer, which gives him a 23% chance of contracting that disease. Also, his risk of lung cancer, if he continues smoking, is sixfold higher than that of other smokers, because he lacks the active form of a certain detoxifying gene that is necessary for the maximum resistance against cigarette smoke carcinogens. In other words, he has a 40% chance of getting lung cancer if he keeps smoking. Now, what is John going to do at this teachable moment?

This information will hopefully be quite useful. He has already decided that he is interested in the information, and his consent to learn the details was informed. People should not be forced to learn this information if they do not want it. However, since he has requested it, there are actions he can take in each of the three situations. For the increased risk for coronary artery disease, a combination of diet and pharmaceutical intervention is prescribed, based upon his particular genetic situation. By knowing what the genetic situation is, risk can be reduced significantly by this treatment.

Regarding the colon cancer risk, John will need a colonoscopy every two years, starting at age forty. That will catch a polyp while it is still small and can be easily removed, and will prevent him from dying from metastatic colon cancer down the road. This is an intervention we know how to do presently; we just don't know how to find those high-risk people like John.

The lung cancer risk has an obvious solution. Of course, John has heard before that smoking is bad for him, but this time he is hearing it in a very personal way: "John, it is bad for everybody, but it's *really* bad for you. You have a forty percent chance of ending up with lung cancer if you do not kick this habit." Although not everyone who hears this will be able or willing to respond, this personal information will heighten the likelihood that John and many others like him will succeed in changing their lifestyles.

While this is a pretty exciting scenario, this is far from the ultimate payoff of genomics. Ideally, by 2015 or 2020, we will see substantially improved therapies for most diseases, so that even if John does get sick, there will be much better solutions to offer him.

Ethical Challenges Ahead

Certain questions have to be addressed if we are going to see a beneficial outcome from this research for the health of all of us. We have to be sure that misuses of this whole set of advances do not eclipse the benefits. If we are not vigilant, something that should have been a wonderful revolution will be turned into something harmful. There are a number of potential worries here, and it would benefit all of us involved in this discussion to keep Proverbs 19:2 in mind: "It is not good to have zeal without knowledge, nor to be hasty and miss the way." Scientists can get quite zealous by saying that what they do is the most important thing on the planet. At the same time, some of the ethical debates have gotten off track and focused on some scenarios that are not very realistic. Zeal for doing God's will and for a good outcome needs to be combined with a clear commitment to understanding all of the intricacies of these issues.

Some questions that should be raised, then, are as follows. First, will we prevent information about people's genes from being used against them in discriminating ways? We ought to be able to say "yes," but cannot as of yet. More concretely, will John, ten years from now, use his genetic knowledge to craft his own health care plan and reduce his risk of getting sick, only to find out that his health insurance has been cancelled because of those very risks he's trying to manage? Will he be fired because he might be a bad health risk? Such outcomes need to be prevented, because they are unjust. We need to take predictive genetic information off the table when it comes to health insurance and employment decisions. Many states have taken action on this, and the federal government has also made some progress, particularly regarding group health plans, but we do not have adequate protections across the board, and we certainly don't have any such protections in employment. In a recent Senate hearing on genetic discrimination in the workplace, it was encouraging to hear both the Democrats and the Republicans stating that this is an important issue, and that something needs to be done about it this year. However, that statement has been made the last three years running, and nothing has happened. People tend to be reluctant to take legislative action before there is a crisis, but it is obvious that the train is coming down the track, and we ought to get ready for it. There is a role not only for preventive medicine but preventive legislation to take care of potential problems before they injure a lot of people.

Second, when is the right time for a genetic test to leave the research lab and move into clinical practice? Will medicine and science take the initiative in this decision? Will it be driven solely by the marketplace? Will we end up with a bunch of tests that aren't properly validated, giving people information that is confusing or even wrong? A federal advisory committee on genetic testing is wrestling with this issue right now. While encouraging this field to blossom is important, we must also encourage some oversight so that these tests are not applied to people prematurely.

Next, there is an educational challenge. Is the general public ready to incorporate genetic information into its medical care? Even if it is ready, it will be a difficult transition. These are complicated issues, couched in terms like statistical risk. Health care professionals are, for the most part, unprepared. Most physicians have not had much training in genetics, but they will be on the front lines as the practitioners of genetic medicine. Science must figure out how to prepare them for this test. The National Human Genome Research Institute has been working quite closely with the American Medical Association and the American Nursing Association to try to do just that.

The prospects for genetic medicine are also complicated by the fact that access to health care is not universal. The United States has a dichotomous health care system. While many have excellent health care, over forty million Americans do not have any health care coverage. New technologies that come along tend to be both very expensive and available only in certain places. Genetic technologies could well drive an even larger wedge between those who have access to good health care and those who do not. Genetics is not the cause of this problem, but it could augment the gross inequality in our health care systems.

A more philosophical concern is whether or not all this genetic knowledge erodes the concept of free will. As talk about "the gene for this" and "the gene for that" becomes household conversation, more and more questions will arise about whether we are victims of our DNA. For instance, *Time* Mag-

15

azine has published an article about the existence of a gene for infidelity. A religious magazine recently ran an essay suggesting that there may be a gene for spirituality and that this is why some people don't go to church. As people are exposed more and more to genetic discoveries, especially people who are not grounded in faith, they may begin to conclude that we are nothing more than machines. This is an erroneous conclusion, as those of us who are well-versed in Scripture know. Science is not going to render free will obsolete; it will not, in fact, shed very much light on free will at all. It will certainly not shed any light on what it means to love someone, what it means to have a spiritual dimension to our existence, nor will it tell us much about the character of God. We must look to other sources to provide light for our spiritual journeys. Science is an incredibly powerful tool for understanding the natural world; but it is poorly designed for understanding these other aspects of who we are.

Perhaps the most difficult question of all is how we are going to set boundaries around the use of genetic technology. First of all, we must dispel some of the false concepts of the abilities of science in the field of genetics. We do not possess the technical ability to create a "designer baby." Many of the scenarios that alarm people are of this nature, where it is suggested that it is possible to tinker with a person's DNA sequence and end up producing an offspring optimized for intelligence, athletic ability, physical attractiveness and the like. Not only does science not understand the genetics of these characteristics, they are also highly affected by environment. In other words, these scenarios often assume sweeping genetic determinism, and we ought not to fall into that fallacy. Furthermore, such genetic tinkering is technically far beyond our present ability.

A strong case can be made, however, for declaring a moratorium on interventions in the human germ line, even for the purpose of treating disease. There are no compelling arguments that the risks one takes in permanently changing hereditary material justify the benefits. And if the risks outweigh the benefits for medical purposes, certainly they do for nonmedical purposes.

There is a way, however, in which people already can shift the odds in the genes their offspring will inherit. Anyone who has seen the movie *Gattaca* has watched a chilling demonstration of this model. It involves using *in vitro* fertilization to produce a large number of embryos from a particular couple and then using preimplantation diagnosis to select the one that has the optimum collection of genetic information. As the smooth-tongued counselor in the film explains to the couple, "We are not doing anything unnatural here; we are just trying to give you a child that was the best you could have created." That exact application of technology is already being utilized for couples who are at high risk for very serious diseases, such as Tay-Sachs. People's views about that will be shaped by their understanding of the status of an embryo, because this process involves decisions to destroy the embryos that will develop certain diseases. Many people are already very disturbed about such destruction.

Sometime in the next few years, this issue will have to be faced. Inevitably, somebody will wish to utilize this technology for something that is not a treatment of a disease but the optimization of traits such as body weight or a personality trait. In doing so, no clear line will have been crossed. There is a spectrum here with a large gray zone. Is preimplantation diagnosis following *in vitro* fertilization something we want to do for enhancement of characteristics? This question immediately brings two sacred principles into conflict: the ability of a couple to make a decision about their own child bearing versus society's interest in making sure we do not go down a pathway that we collectively regret. In our pluralistic society, it may not be realistic to expect a consensus. Nevertheless, the church needs to take a stand on this issue. Addressing such a focused and scientifically plausible scenario creates a higher likelihood that the church will be heard than if it is speaking to a more diffuse set of hypothetical concerns.

The issues and struggles that we are facing now make this a difficult time. However, the worst thing to do at this point would be to turn around and go back. Winston Churchill expressed this principle very well when he said, "If you are going through hell, keep going." It's vitally important that we persevere, particularly during this time of crucial decision making. Edmund Burke suggests another important tenet in this circumstance: "All that is necessary for evil to triumph over good is for good men to do nothing." Men and women of principle must take a stand here; fading into the background and waiting during this time could be dangerous.

A final thought is found in the reflections of Copernicus. It is really a defense of science in general, and speaks to those who think that there are areas of God's creation which He did not intend us to explore. Copernicus wrote: "To know the mighty works of God, to comprehend His wisdom and majesty and power, to appreciate in degree the wonderful working of His laws, surely all this must be a pleasing and acceptable mode of worship to the Most High, to whom ignorance could not be more grateful than knowledge." God gave us an intellect, He gave us curiosity, and He hoped we would use it to try to understand His creation. When we have a chance to glimpse something that God knew before but no human did, it is a wonderful feeling and something I believe He approves of. After all, God is the greatest scientist of all. At the same time, those explorations must be carried out with a great sense of responsibility about how scientific knowledge, which is neither good nor evil in and of itself, is applied. That is where the moral stakes lie.

Xenotransplantation

E. DAVID COOK

One of the more recent arrivals on the biotech scene is the science of xenotransplantation. Xenotransplantation is defined as the use of live cells, tissues, or organs from a nonhuman, animal source transplanted or implanted into a human or used for contact with human body fluids, cells, tissues, or organs that are subsequently given to a human subject. It is vital to locate the development of xenotransplantation in the context of the overall goals of medicine, the rise of biotechnology, and the success of transplantation.

The Context

There are many formulations and expressions of the goals of medicine. Some would say that medicine exists to prevent disease and to promote health, while others see its essence as relief of pain and suffering. Some would summarize medicine's goals as to care and to cure: when no cure is possible, then care is still the focus. Still others understand medicine in terms of avoiding premature death and trying to ensure a peaceful death.

Modern biotechnology boasts numerous accomplishments. In reproductive medicine, we have been able to produce "test-tube" babies and develop egg and sperm donation. Surrogacy and postmenopausal childbearing are both possible. Very premature babies can be kept alive and helped to develop far beyond the expectations of even five years ago. We are also able to transplant human hearts, lungs, kidneys, livers, corneas, and skin. We can take brain cells from aborted fetuses and from animals and implant them into human brains. In the area of genetics, we are seeing a dramatic growth in understanding the

18

links between genetics and disease both in terms of predisposition and in actual incidence of disease from genetic causes. Such understanding is paving the way for genetic replacement, genetic modification, and cloning.

After a faltering start, human to human transplantation (allotransplantation) has become part of regular, accepted treatment and the success rate of such transplants has grown considerably both in terms of life expectancy and quality of life achieved. In kidney transplants, we can expect that four out of five will last at least one year, and that two out of every three will last at least 5 years. The success rate is growing for liver transplants; recent statistics show that half will last at least five years. All of this adds to both length and quality of life, evidenced by lives which have greater freedom from regular medical procedures, as well as greater freedom in travel, exercise, lifestyle, eating, and drinking. Such transplantation is also cost effective. In the United Kingdom in the late 1990s, a kidney transplant cost approximately £10,000 for the operation and £3,000 a year to maintain the patient. In contrast, dialysis cost £18,000 per year.

When we consider the question of xenotransplantation, we are not dealing with a completely new technology. Already, inert heart valves from pigs are used in heart valve replacement operations. Tissue for bones and skin is grown and developed from pigs. There are many relatively successful transplants of pancreatic islets from pigs into humans to help those with diabetes. There are also experiments which take brain cells from pigs and implant them into people suffering from Alzheimer's disease.

Rationales for Xenotransplantation

One of the unforeseen and unintended consequences of legislation requiring the wearing of seat belts in cars and buses has been a decrease in the number of organs available for transplantation. In the UK in the late 1990s, over 5,000 people were waiting for kidney transplants. In the USA, there are over 400,000 people waiting for organs for transplant. If we were able to extend the waiting lists to reflect the need in the world as a whole, we would find an immense number of people awaiting every kind of organ. This situation has caused some people to examine the feasibility of transplanting animal organs into humans. Quite properly, those who object to xenotransplantation argue that there are alternative, human sources for organs. There are, however, some problems with relying exclusively on human organs.

One of the general issues which faces transplantation from human to human is that of how we define death. Once we argued that if someone stopped breathing, they were dead. Then we began defining death by heart stoppage, until we found that breathing and heart functions could be restored. Brain death was next used as the criterion, and increasingly brain stem death is the most reliable and widely accepted criterion of establishing when someone is dead.

The use of living organ donors instead of cadavers is one way to avoid questions of the definition of death, but it can create its own moral dilemmas. For instance, if a living kidney donor's remaining kidney fails, should that person move to the top of the waiting list? One answer, designated the "Exeter" protocol in the UK, was a system of elective ventilation for dying patients. With personal and familial consent, they would be moved to intensive care and ventilated in order to keep their organs in good shape for transplantation. However, this creates further dilemmas: a) the use of such limited, expensive resources, b) what to do if the patients begin to improve because of the ventilation, and c) the danger that the consent is a result of societal pressure to misuse people as merely a means to an end.

In parts of Europe, there has been a strategy of organ donors opting out of donation programs rather than the traditional US and UK system whereby individuals opt in to donate their organs. Some, including the British Medical Association, have begun to argue for a presumed consent for organ removal and donation. This raises difficult questions about the "ownership" of organs and bodies, especially after death.

Many public steps have been taken to encourage more people to donate their organs. In spite of alleged public support, when it actually comes to completing organ donor cards or ticking the appropriate box on a driver's license form, the public has been consistently unwilling to participate in organ donation.

Researchers have attempted to develop artificial organs. This has been more successful in the area of the heart than with other organs. However, even in heart transplantation artificial organs are not without considerable problems. Chronic dependence on life support with no possibility of existence separate from it is hardly a satisfying option.

In light of the vast numbers worldwide who would benefit from xenotransplantation and the need for associated drug regimes, sacrificial animals, and animal breeding and housing facilities, the opportunity for financial gain is astronomic. This has been a key factor in the critical focus on such biotechnology and the money spent on its research. The concern is that the desire for profit may interfere with or control research and its results.

Given the possibility of organ rejection and the need for immuno-suppression, scientists have begun the genetic modification and even cloning of animals, particularly pigs. Such animals must be among the most pampered animals in the world. They live in a pathogen-free, air-filtered and airconditioned facility. They are fed on a special diet. They see the vet on a regular basis. They are given toys for stimulus and their pen is carefully constructed to prevent injury. All that is required is that when and if the time comes, they will lay down their lives for human beings.

Problems With Xenotransplantation

The main concern among the scientific community in examining xenotransplantation has been the threat from Porcine Endogenous Retro Viruses (PERVs).[1] We have seen the kinds of dangers that animal to human disease transmission can create with Bovine Spongiform Encephalopathy (BSE or "Mad Cow Disease"), not to mention such transmission being one of the alleged causes of HIV and AIDS. Such medical problems have made researchers rightly cautious about proceeding with animal to human transplantation until and unless there is general consensus that PERVs will not release some kind of "plague" and that adequate surveillance and biosecurity is in place to control any and every unforeseen outbreak.

A secondary, but significant, concern is the emotional response which may or may not be associated with xenotransplantation. The "yuk" factor plays a significant role in the public's response to all kinds of biotechnological developments. The basis of such an emotional reaction may be some deeply-seated human response to going beyond proper natural limits or a mere "feeling" that this is one step too far. Certainly the "yuk" factor has been key in the limiting of some biotechnological processes, such as in some countries that have banned the use of eggs from aborted female fetuses in the development of live births. A case in point is the UK where the Human Fertilization and Embryology Authority received many letters from the public, who wrote in expressing their disgust with such methods.

Yet another concern has to do with psychological reactions: first, the reaction of those who are recipients of an animal organ; second, the response of others to those who are recipients of such organs. Will this lead to a higher

---

1. C. Patience, Y. Takeuchi, R. A. Weiss, "Infection of human cells by an endogenous retrovirus of pigs," *Nature Medicine* 3 (1997): 282-286; C. Patience, Y. Takeuchi, R. A. Weiss, et al., "No evidence of pig DNA or retroviral infection in patients with short-term extracorporeal connection to pig kidneys," *Lancet* 352 (1998): 699-701; W. Heineine, et al., "No evidence of infection with porcine endogenous retrovirus in recipients of porcine islet-cell xenografts," *Lancet* 352 (1998): 695-698.

rate of rejection simply based on psychological reaction, rather than any specific physiological ground? This question is yet to be answered, but it is a valid one which must be considered.

The other foundational concern of those who oppose xenotransplantation, apart from safety concerns, has to do with animal issues. At the heart of this debate lie some crucial questions regarding the nature and status of animals. For instance, it is clear that there is a wide range of capacity and ability among animal species. In terms of intelligence, the ability to communicate and the capacity for and exercise of some kind of social life, there is great diversity. In observing members of various species of the monkey and ape groups, there is evidence of all three of these abilities. There is also research which claims that pigs are highly intelligent animals and participate in a clearly definable social life. However, such observation still leaves unanswered the question of how far it is appropriate to go in using such species for research and for xenotransplantation.

What cannot be disputed is that animals have a clear and established capacity to suffer and feel pain. But there is always a danger of what is called the "pathetic" fallacy. This warns of the assumption that animals feel exactly what human beings feel. Even if we agree that animals have some or all of the capacities under dispute, there remains the question of whether or not preference in terms of care and survival lies with the human being rather than the animal. In answering this question, the close affinity of primates with human beings has made many cautious about any kind of use of primates in medical experimentation. This may also be because of the risk of species extinction and the nature and extent of disease among primate populations.

Peter Singer has been the main opponent of what he terms "specieism" (the preference human beings show for their own species at the expense of other animal species). He argues that, in terms of capacities, many humans are in fact inferior to animals rather than superior, as specieism suggests. Many animals, Singer argues, are intellectually superior to human embryos, early infants, anencephalic babies, people in a Persistent Vegetative State (PVS) and demented persons. Such a view demotes the notion of personhood — not to mention spirituality — to a view based only on certain physical or mental capacities.

Underlying all these concerns about and for animals is the moral question of whether animals have rights. The Kennedy Commission in the UK argued that it was better to talk in terms of the interests of animals than of animal rights. All animals have interests which need to be safeguarded, and not just species threatened with extinction. There needs to be a basic concern for animals, only in part because they are used to play a key role in the preservation of human life and well-being. Such concern is a matter of good steward-ship of the world.

In many circles, there is a distinct unease that in undertaking xeno-transplantation, human beings are going a step too far. Instead of respecting and following nature, they say, xenotransplantation crosses fundamental spe-cies lines drawn by Nature itself (or God). The problem with this argument is that medicine is then accused of "playing God" as soon as it interferes in any natural process. Farming and agricultural methods seem to have been cross-ing the species lines for years without qualm and without negative results or resistance on the part of these same circles of concern.

It will be impossible to ensure that there are no negative consequences or risks at all in xenotransplantation or in any medical intervention. While medicine is a science, it is never a foolproof science with one hundred percent certainty. There are too many variables which cannot be controlled. This does not mean that scientists are careless, but that there will and must come a point where the risk taken is considered as small as possible and the likely benefit so great that some degree of risk can be justified.

Xenotransplantation provides one extreme measure of hope for those who face a premature death. As such, there is the possibility of a clear and real benefit to the individual. Accordingly, issues of fair access for all individuals need to be addressed. On the other side, there has to be balanced a concern for the community as a whole. It is unfair and wrong to put the well-being of one individual before the well-being of the majority. There needs to be a care-ful balancing between the benefit to the individual and the risk to the com-munity. At present, the balance is quite clearly leaning toward the need to protect the community. As more evidence is gathered which shows that the risk to the community is negligible, the balance may well move from the com-munity to the benefit of the individual.

With concern about the financial drive for profits, there must be a clearly based, solid scientific ground before xenotransplantation is permitted. Com-mercial interest may well be exercising too great a degree of selection and con-trol over the work of science. However, it is interesting to note that the members of the Roslin Institute (creators of Dolly, the cloned sheep) are abandoning xenotransplantation research for commercial reasons.[2] With the withdrawal of central governmental funding for much research in the UK, commercial fund-ing has become the norm and order of the day. Such funding and research need to be controlled by external objective bodies and not just the financial benefi-ciaries in order to keep scientific medical research in an objective range.

2. A. Coghland, "It's just business, says Roslin," *New Scientist* 19 (August 2000): 5.

Crafting a Way Forward

The role of the media in reflecting and creating public opinion is very great. The way that the media deal with medical issues (especially cutting-edge medical technology) will be crucial for the acceptance, permission, or rejection of xenotransplantation. It is important that a balanced, realistic perspective is presented to the public.[3] Advocates on both sides of the argument will do well to win the hearts and minds of public opinion through wise use of the media. This will foster the attitudes and perceptions of transplantation in all its forms and the degree of acceptance or rejection that society will express.

Recently, the Council of Europe Working Party on Xenotransplantation has produced a draft report which includes a consideration of the cultural, ethical, and religious aspects of xenotransplantation.[4] Different cultures and religions have very different views of the nature and status of animals. Some reject the eating of all animals. Some limit the kinds of animals which may be consumed. A few allow almost indiscriminate use of animals for food. It is vital that cultural and religious sensibilies are carefully considered in discussions and decisions about proceeding with xenotransplantation. It is interesting that even those religions which are opposed to the eating of pigs are not necessarily opposed to saving human life by means of xenotransplantation. It seems that other than the means of absorption, there may be little actual difference between eating a bacon sandwich and having a pig's kidney transplanted.

Consent and patient autonomy lie at the heart of modern medicine. Nothing should be done to a patient without explicit consent except in emergency situations and that for life-saving purposes alone. The high standard of fully informed, valid, and voluntary consent given by a rationally capable person is a core value in medical treatment and research. But in the case, certainly initially, of the first xenotransplantation patients, such standards of full information and coercion may be almost impossible. We will not know all the likely consequences and outcomes of a particular xenotransplant. We will not be able to give a full account of the possible side effects and consequences of such a procedure. The first patients are likely to be those who are in an extreme situation, where all other possibilities have

3. David Cooper and Robert Lanza provide one such presentation in their related work, *Xeno: The Promise of Transplanting Animal Organs into Humans* (New York: Oxford University Press, 2000).

4. Report from the Council of Europe Working Party on Xenotransplantation to the Steering Committee on Bioethics (CDBI) and European Health Committee (CDSP), 2000-2001.

been exhausted. It is hard to see that appropriate consent can be fully given under such circumstances.

Like consent, confidentiality is a core value in modern medicine and doctor-patient relationships. However, given the potential threat to the general public, the conditions of surveillance will be extremely strict and cautious for the first xenotransplants. It will be necessary to know the details of the patient and his or her "close contacts." They will have to agree to be followed up and tested regularly, even postmortem, to ensure that there are no or limited untoward effects. To guarantee such protection of the public, there will be little that is confidential. This seems to go against the trend in medicine.

Others have argued that it is possible for an individual and his or her family to waive the normal guarantees of consent and confidentiality in extreme life-threatening situations. It is perfectly proper for someone to consent to have his or her confidentiality and need for fully informed consent waived if the time and extreme danger of the situation so warrant. Nevertheless, in such situations, the importance of the ethical standards involved still dictates that as much confidentiality and as fully informed a consent as possible be preserved.

Biosecurity arrangements are yet another exceedingly important practical aspect of this process. If and when xenotransplantation becomes accepted procedure, protections against intentional or unintentional biological harm caused by it will need to be already in existence. It is no accident that strategies including these are already being played out in every country that is considering the regulation of xenotransplantation. However, in the midst of this, it is important to ask whether such protections can be absolutely guaranteed. Although such an assurance is not realistic, that should not stop us from striving for the very best biosecurity possible.

In the first stages of xenotransplantation, it will be vital to monitor all patients and their close contacts regularly. Initially it seems that such surveillance and monitoring will be lifelong. It will require regular testing, and the possibility of segregation and isolation cannot be ruled out. After death, postmortem investigation will be necessary to discover what results may have followed from the xenotransplant. With the passage of time and the increased certainty that there is minimum risk, such extreme requirements for surveillance can be eased. Cross-border, international agreements and procedures will need to be in place to preserve the well-being of all who may be affected by xenotransplantation. That may well include all of us.[5]

5. For a review of ethical issues involved in the first clinical trials of xenotransplantation, see S. Welin, "Starting clinical trials of xenotransplantation — reflections on the ethics of the early phase," *Journal of Medical Ethics* 26 (2000): 231-236.

In the desire to help the progress and development of medical science and to save human life, we need to realize that moral questions must be addressed. Just because we have the technology to do something, does not mean that there is some kind of imperative that we must use that technology. Just because we can do something, does not mean that we ought to do it. We need to look carefully at the technology itself, the motives for its use and development, the nature of what is involved in using that technology, and the likely consequences which may follow its use. Only then can we properly allow the practice of a technology.

Medicine exists in order to help preserve the life and well-being of human beings. At times it seems that modern medicine is determined to preserve life by all means and at all costs. This drive should cause us to reflect on what we really mean by sanctity of life and whether we are in danger of inappropriately trying to resist death.

Human beings, science, and medicine are extremely limited. The success of medicine at times has given an impression that life can be endless, that we can all be one hundred percent healthy and that science will solve all our medical and other problems. None of that is true. The question is what limits we will live with and how we come to terms with the reality that the certainty of death is one hundred percent and that none of us is or will be a perfectly well and completely healthy individual.

Throughout Europe, the USA, Canada, and Australia there are moves to regulate and control xenotransplantation.[6] National authorities and bodies like the United Kingdom Xenotransplantation Interim Regulatory Authority (UKXIRA) and the Federal Drug Administration (FDA) and the Centre for Disease Control (CDC) have all been involved in cosponsoring work on how xenotransplantation research must be regulated and controlled.[7] UKXIRA publishes an annual report on xenotransplantation and has also published proposed standards for surveillance and biosecurity, which are available via its website.[8] Awareness of the potential risks of and from xenotransplantation have galvanized countries together and individually to look at measures for the control of xenotransplantation and the facilities, monitoring and surveillance of the patients, families, medical nursing, and veterinarians, and animals involved.

Underlying all these standards set by national and international bodies

6. The Canadian work can be seen on the Internet at: http:// www.hc-sc.gc.ca/hpb-dgps/therapeut/htmleng/btox.html.

7. All these organizations can be accessed via the Internet: UKXIRA at http:// www.doh.gov.uk/ukxira.htm; FDA at http://www.fda.gov; CDC at http://www.cdc.gov.

8. Ibid.

are fundamental values stemming directly from the Judeo-Christian and Hippocratic traditions. For Christians, defending both the community and the individual from harm is a basic value that sets limits even to the doing of good. The biblical standard of responsibility before and answerability to God runs from the Garden of Eden to Judgment Day. It underlies the responsibility of scientists and researchers not only to their own consciences, their professional bodies, or to regulatory authorities, but also to the Creator of all knowledge and understanding. The close parallels between the concerns of medical research and Judeo-Christian values should give Christians hope that we can influence and affect even a secularized society in the God-given role of restraining evil and reinforcing good.

Christian Perspectives

What perspectives do Christians have to offer in the area of xenotransplantation? To answer this question, we need to touch upon such matters as pluralism, natural law, human rights, and a proper view of life and death.

It is a truism to say that we live in a pluralist world. That means both that we are confronted with a variety of moralities, religions, and worldviews and that there is a philosophy that stresses the need for that variety and the benefits which follow from it. Christians are in the business of truth and believe not only that God is the Source and Author of all truth, but also that Jesus is the Way, the Truth, and the Life. How then are Christians to live and function in a world which does not share and indeed challenges fundamentally that perspective and outlook?

Christians have nothing to fear from science and the pursuit of truth. Believing that all truth is God's truth, we can be confident that science properly conceived and practiced will lead us back to God. That still leaves open the question of how Christians are to act and behave in a world of difference in outlook. The New Testament Church grew and developed in a pluralistic, multi-faith, multi-morality context. The New Testament offers a pattern of how to present the gospel and live in such contexts.

Jesus taught in the Sermon on the Mount that Christians are to act as salt and light in the world.[9] As salt they are to preserve what is good and prevent decay. They are to ensure the wholeness of what it means to be human. They are to help women and men to grow and flourish and they are to bring healing. As light, Christians are to bring life, to show up the nature of and

---

9. The Sermon on the Mount is found in Matthew 5–7 and Luke 6:20-49.

27

danger from evil and to show humanity how to live. Jesus Christ, the man for others, was the perfect embodiment of the Christian life. Paul in Romans 13 sets out the role of government and those in authority. They are there in order to punish evil and to reward good. Christians, therefore, must be in the business of restraining evil and reinforcing good. That means that as new technologies come along we are to hold them accountable to criteria promoting the restraint of evil and reinforcement of good.

That will certainly mean exploring the common ground we share with all human beings made in the image of their heavenly father, whether they realize it or not. All are subject to the common grace of God, which does not leave anyone without some kind of moral sense. As Christians we can appeal to and build on that common ground, which must surely relate to human flourishing and the avoiding of human harm.

One of the tragedies of Protestantism has been a failure to explore the nature and meaning of natural law. If we believe that God is the Creator and sustainer of the world, then there is a way in which His standards are written in the laws and book of nature. It is not easy in a fallen world, as fallen men and women, to recognize God's natural law and original creation intention, but that does not mean that we can't have ideas of what is good and what is bad for humanity.

We must also beware of treating nature just as a means to our own ends rather than as a gift from God. We are stewards of nature and have a responsibility to care not only for the world itself but also for the created order and to give it an appropriate respect.

In the Eden setting, humanity is given responsibility to name the animals, to care for them and to act as a steward.[10] In Genesis 9, after the flood, for the first time human beings are given permission to use animals as food. We are entitled to eat to live, and animals are part of the freedom. Yet there are very careful dietary laws set forth in the Old Testament in order to ensure public health and well-being. While the New Testament indicates that specific dietary laws are not binding *per se* on people today, the issue underlying them is still a legitimate concern: we should look very carefully at how we use animals and which animals we use. We should be particularly careful about how far it is appropriate to use animals for any kind of research.

Some have argued that God has set the species each in their own place and species lines ought not to be crossed. This has never been the central view of the Church in response to agricultural, horticultural, and animal developments. We do not take Old Testament laws and apply them without looking

10. Genesis 1 and 2.

carefully at the point behind the law and asking how that point can be safe-guarded and applied in today's world.

Another related concern is the development of the notion of human rights, which is found full flowering in the Western liberal tradition. While the core concern of this movement to respect and protect human beings is generally in harmony with Christian values, we need to be critical before we accept the idea of the nature, foundation, and extent of human rights. The Bible talks of duties and responsibilities; thus any talk of applying rights to animals or even propounding human rights as some overarching moral base needs careful thought and scrutiny. In medical technology we have specific duties and responsibilities to patients, their families, the public, animals, and to science.

Ultimately, a Christian perspective on xenotransplantation must locate all of these various ethical questions within a proper understanding of death and dying. Medicine has been able to delay death but there is and has been considerable cost in that. Technology can dehumanize and depersonalize people, especially at the end of life. We need to ensure that we keep death in perspective. That certainly means that death is not the worst thing that can happen to anyone and need not be resisted at all costs and by all means. On the other hand, it does not mean that death is to be meekly accepted. There is a sense in which death is an enemy — the last enemy — that is to be fought and resisted. There is another sense in which death is part of this life and merely a stage in the transformation from one degree of glory to another.

Christians need to be clear about their attitude, not only to their own death and the life beyond, but also to how our society handles death, those who are dying and those who remain after the death of a loved one. Modern medicine is often the battlefield in dealing with different perspectives on death. Christians in their living and dying, their care of the dying, and their presentation and living out of their eternal hope have much to offer to a world which fears death and the process of dying, perhaps above all else.

Ultimately, then, from a Christian perspective, it would appear that there is no overwhelming moral objection to the technique in and of itself, unless we wish to reject any and every form of species and genetic manipulation. This would be hard to do consistently, given the developments of plants, animals, agriculture, and farming over the centuries. Nevertheless, Christians have plenty to say about xenotransplantation. As those concerned to act as careful watchpersons to warn society of the danger of making evil out of good or taking a technology and technique too far, Christians must ensure that humankind is a responsible steward, before the living God, over the animal kingdom and the science and knowledge entrusted to us. We must beware of putting humanity at the center and seeking to wrest God's sovereign role by placing ourselves and our well-being before all else. Perhaps most of all, Christians need to help women and men live with the limits of our human frailty and humanity. We shall all die and death is not to be resisted by all means and at all costs. There is, after all, an eternal life beyond.

Transgenics

NANCY L. JONES AND LINDA BEVINGTON

Transgenic animals are animals which have had DNA from another species inserted into their genetic code. The goal of this type of transgenics is to produce a hybrid animal that is able to pass on genetic material from two different species to the next generation. Inserting genes from one species into another species to create a transgenic animal is considered the most powerful technology for modeling disease processes and for determining the mechanisms by which genes are regulated during development. Transgenic animals, also called "bioreactors," allow the effects of various factors on a gene's function to be tested in a whole animal rather than merely in a test tube or cell. By inserting human DNA into an animal such as a mouse, medical researchers are provided with important information which may help them in their efforts to conquer human disease. Transgenic technology has undergone explosive growth in the last decade. A 1989 search of the NIH Computer Retrieval of Information on Scientific Projects (CRISP) database for government-funded human/animal transgenic research revealed only twenty-one grants — a number which grew exponentially to 1,820 grants by 1999. Today nearly twenty percent of government-funded research grants go toward underwriting transgenics research.[1] This chapter will focus primarily on human/animal transgenics, but it will also include some information on transgenics involving only nonhuman forms of life in order to clarify the broader context within which this issue arises.

Since 1980, mice have been genetically engineered to make "knock-out"

---

1. Computer Retrieval of Information on Scientific Projects (CRISP) allows searches of all NIH-funded grants (http://www.commons.cit.nih.gov/crisp/).

mice models, where a single gene is rendered nonfunctional.[2] Such models are designed to help researchers determine the function of a particular gene. Techniques were later developed to replace the knocked-out genes with functional genes from other species. These are called "knock-in" mice. Until recently, most transgenic animals were created by inserting just one or two genes from one species into an animal of another species. The technology was limited by two features: the amount of genetic material that could be transferred (usually only a single gene, 40-50 kilo bases) and the ability to locate the gene precisely so that it was functional.[3] Therefore, transgenic mice typically had only one or two foreign (human) genes.[4]

The current trend is to insert more and more human DNA into an animal of another species. Newer techniques using yeast artificial chromosomes (YACs) and bacterial artificial chromosomes (BACs) allow insertions of up to one third of a chromosome (or 500-1,000 genes) to create a transgenic animal.[5] YAC and BAC techniques facilitate transfer of multiple genes and other cellular elements required for the faithful regulation of a gene. YAC technology has been used to create transgenic mice and has been applied to farm animals like rabbits and pigs.[6] This YAC transgenic technology is currently being employed to create transgenic pigs for the purpose of developing organs for human transplantation (a technique known as "xenotransplantation" — see also David Cook's chapter in this volume).[7] The successful application of these techniques has raised some old questions surrounding animal modeling for human disease, as well as important new ones about the ethics of such options.

2. J. W. Gordon, et al., "Genetic Transformation of Mouse Embryos by Microinjection of Purified DNA," *Proc Natl Acad Sci U S A* 77 (1980): 7380-84.

3. J. W. Gordon, G. Harold, and Y. Leila, "Transgenic Animal Methodologies and Their Applications," *Hum Cell* 6 (1993): 161-9. The human genome is 3×10^6 or 60,000 times larger than 50 kilo bases.

4. Mice are the most widely genetically engineered animal. Recently methodologies have been developed to extend transgenic technology to other common laboratory animals and livestock.

5. It would take 1,500 YACs to cover all the human genome. See K. R. Peterson, "Production and Analysis of Transgenic Mice Containing Yeast Artificial Chromosomes," *Genet Eng* 19 (1997): 235-55.

6. G. Brem, et al., "YAC Transgenesis in Farm Animals: Rescue of Albinism in Rabbits," *Mol Reprod Dev* 44 (1996): 56-62. R. E. Hammer, et al., "Production of Transgenic Rabbits, Sheep and Pigs by Microinjection," *Nature* 315 (1985): 680-3.

7. Some transgenic pigs expressing human complement regulators have already been made with YAC technology. G. A. Langford, et al., "Production of Transgenic Pigs for Human Regulators of Complement Activation Using YAC Technology," *Transplant Proc* 28 (1996): 862-3.

Ethical Questions

First of all, the concept of genetically manipulating animals so that they become diseased and defective distresses much of the general public. For example, that mice are being genetically manipulated to develop cancer causes concern among many. Animal rights proponents have claimed that patenting will increase the suffering of experimental animals because the unusual experimentation raises the prospect of considerable gain.[8] However, the use of animals to model a disease process has long been a standard tool of medical research. The medical research community has been purposefully creating disease states in animals by administering drugs and selectively breeding for decades, and such methods generally have been met with public acceptance.

Several biblical passages support this type of medical research. First, medical research is an extension of the mandate in Matthew 9:35 to have compassion on those in need and to help in the healing mission. Additionally, people have been given dominion over nature and every living thing. The use of animal models for conquering human disease falls under the dominion and stewardship we have been given over the created order (Genesis 1:26). As such, careful scientific study of animals can be ethically justified. The following standards currently in place for animal research are consistent with an ethical use of animals: testable hypotheses and experimental procedures are peer-reviewed; the studies are demonstrated to be well grounded via cell culture and literature reviews; justification is provided for the appropriateness of using certain animals rather than any other methodology; animals undergo minimal suffering; and the experiments are well planned so that as few animals as possible are involved.

Although the use of animal models for medical research has already been justified and accepted by society, transgenic technology raises new concerns. First, research on mice — the animal used most commonly in transgenics — is subject to less restrictions and regulations than is research on other laboratory animals.[9] Second, genetic engineering of animals which

---

8. L. J. Raines, "The Mouse That Roared," *Biotechnology* 16 (1991): 335-45.

9. Mice are outside the purview of the Animal Welfare Act of 1966 (Health Research Extension Act of 1985, Public Law 99-158, November 20, 1985, Section 495 U.S.C.§§2131-2156, P.L. 89-544, August 24, 1966, P.L. 91-279 & P.L. 99-198). But almost all institutions seek accreditation by the Association for Assessment and Accreditation of Laboratory Animal Care International (AAALAC). This means that they comply with the Public Health Services (PHS) Policy on Humane Care and Use of Vertebrate Animals and institute programs consistent with the Guide for the Care and Use of Laboratory Animals produced by the National Research Council.

involves the insertion and deletion of genes increases the likelihood that the animals will die. The common use of mice as a laboratory tool, coupled with the fact that researchers are becoming increasingly enamored with transgenic technology, has sometimes resulted in the creation of transgenic animals without first developing well-substantiated hypotheses about genetic links to a disease or development disorder.

Other ethical questions that arise from transgenic technology are arguments related to genetic engineering. Overall, there have been few concerns voiced regarding the widespread use of transgenic animals in the United States. In Europe, concern over transgenic animals has focused on the breach of species barriers and the violation of species integrity entailed by the creation of such animals. People are worried about the possibility that transgenic animals might escape and threaten the species integrity of wild animals. Another concern is "zoonotic transmission" of disease, which would create pathogens that cross the traditional species barriers for disease transmission. Genetically-engineered animals could become infected with typical pathogens which might then mutate, thereby gaining the ability to cross the normal species barriers by now recognizing the species from which some of the inserted foreign DNA came. To date, zoonotic transmission has occurred only when the animal was engineered with the specific intention that pathogens would be transmitted across species. Such genetic engineering is done for the purpose of creating an animal model for studying a human disease. For example, the natural species barrier of mice to hepatitis B infection was overcome by making transgenic mice which contained hepatitis B virus genes.[10] Another example is the SCID (severe combined immunodeficiency) mice, which provide a model for studying AIDS. These mice were genetically engineered to possess a human immune system and then were infected with the AIDS virus.[11] Such a technique might be used to create pathogens for biological warfare.[12] However, all data gathered so far suggests that zoonotic transmission of disease results from intentionally engineering animals toward this end, and not from mere happenstance.

Another risk of genetic engineering that arises in transgenics is the way that genetic engineering increases the tendency toward genetic uniformity in agriculture. European and Third World countries have expressed more con-

10. J. W. Gordon, "Transgenic Technology and Laboratory Animal Science," *ILAR* 38 (1997): 32-40.

11. P. Lusso, et al., "Expanded HIV-1 Cellular Tropism by Phenotypic Mixing with Murine Endogenous Retroviruses," *Science* 247 (1990): 848-52.

12. B. E. Rollin, "Bad Ethics, Good Ethics and the Genetic Engineering of Animals in Agriculture," *J Anim Sci* 74 (1996): 535-41.

cern about this issue than has the United States. International concerns include domination by a few companies that would control propagation of livestock and produce according to a particular genetic/transgenic pattern. This might negatively affect the economy of developing nations where farmers often plant subsequent crops with seeds from the preceding crop, since agricultural companies are engineering crops to produce nonproductive seeds. Moreover, what if a pathogen arose that destroyed a transgenically engineered animal or crop? Would world famine result because a large portion of an animal population or crop was from a single genetic strain?

There are other ethical questions in transgenics that need further attention. Some have to do with the potentially harmful outcomes of this research. For example, might the use of transgenics in agriculture lead to the ingestion of genetically-inserted proteins which could thereby cause food allergies or immunological reactions? Other ethical questions involve control over the results of transgenic research. Ethical debate over transgenics in the United States, for instance, has focused on animal patenting. The United States Supreme Court has held that "man-made" living organisms are patentable subject materials — a ruling that is still controversial today.[13] The issues involved in owning (i.e., being able to deny others access to) an entire life form only increase in complexity as more and more human genetic material is transferred into animals.

A Christian Perspective

From a Christian perspective, the various ethical issues discussed above are important. At the same time, a Christian analysis of transgenics should also take into account a more basic principle: the divine created order. The Bible tells us that God designed procreation so that plants, animals, and humans always reproduce after their own *kind* or *seed*.[14] In the biblical view, then, species integrity is instituted by God rather than by arbitrary or evolutionary forces. This notion underlies the Levitical mandates to keep seeds pure and to prohibit cross-pollination and crossbreeding.[15] Christians involved in and/or

13. Diamond v Chakrabarty 447 U.S. 303 (1980).
14. Gen. 1:11-12 refers to seed-bearing plants and trees created "according to their various kinds," Gen. 1:21 to "creatures of the sea and every living and moving thing . . . according to their kinds."
15. Lev. 19:19: "Keep my decrees. Do not mate different kinds of animals. Do not plant your field with two kinds of seed." Deut. 22:9: "Do not plant two kinds of seed in your vineyard; if you do, not only the crops you plant but also the fruit of the vineyard will be defiled" (NIV).

concerned about transgenics should seek to determine whether the creation of a human/animal hybrid violates the biblical notion of species integrity.

The complete fusion of human and animal genomes via the union of sperm and egg from different species runs counter to the sacredness of human life as created in the image of God. Biblically, bestiality (sex between humans and animals) is forbidden and punishable by death.[16] Some might assume that the severity of this penalty was due to the defilement of the physical body or the "heart" (that is, a question of means), rather than to a concern about the creation of viable offspring with the genes of two species (the ends). Certainly, God is concerned with our activities — with the means we employ.

However, the ends are important too. The question remains as to the acceptability of the laboratory creation of transgenic offspring, where no physical copulation occurs. Is the simple fact that the resulting offspring have genetic material from different species sufficient to render transgenics unethical? Closer examination of the Bible suggests that Scripture *is* concerned with more than just the physical defilement associated with sex between a human and an animal. Leviticus 18:23 ends with the phrase "it is confusion." The word confusion *(tebel)* means "in violation of nature or divine order." This word is used in only two contexts: when a woman lies with a beast and when a man lies with his daughter-in-law. The second example could, of course, result in viable offspring. Additionally, the word *raba'* — translated in these passages as "mate, gender or lie down" — has the inherent meaning of "copulation" or "breeding." Breeding, too, suggests the potential for offspring. These passages leave us with the suggestion that by divine order all things should reproduce only after their own kind — that interspecies "reproduction" (a misnomer in itself), especially involving humans and animals, is prohibited.

We must now ask whether inserting, for example, an insulin gene from a human into a pig would violate the divine order. To address this question, one must first determine if there is a significant difference between genes from diverse species and if those genes have the same function. In the field of molecular biology, individual genes are classified primarily by their function. Such a classification highlights the homology, or similarity, of the DNA se-

16. Lev. 20:15-16: "If a man has sexual relations with an animal, he must be put to death, and you must kill the animal. If a woman approaches an animal to have sexual relations with it, kill both the woman and the animal. They must be put to death; their blood will be on their own heads." Lev. 18:23: "Neither shalt thou lie with any beast to defile thyself therewith: neither shall any woman stand before a beast to *lie down* thereto: *it is confusion*" (KJV).

quences of genes which have the same function but are from different species. The primacy of function over species may explain why the scientific community in the United States hasn't felt the need to justify the technique of inserting human genes into animals. A gene would be a gene no matter what species it was obtained from. However, genes coding for the same function may in fact differ in various degrees among species. These interspecies differences can have dramatic effects on the function of a single gene or on the interplay between that gene and other genes. Transgenic animals have shown that although a gene may code for a protein with a particular function in one species, the expression of that protein in a new host species can have a very different effect. In other words, a gene is *not* just a gene regardless of what species it comes from.

The next question to be asked is whether the insertion of a single human gene into another species could cause observable changes in the resultant transgenic animal. To date, transgenic research has shown that it is unlikely that the insertion of a single gene from one species into an animal of another species would change the animal's phenotype. For example, the insertion of a single human gene into a mouse would not be expected to produce an observable human characteristic. The phenomenon known as pleiotropy (in which one gene and its product controls or codes for more than one trait by turning on or off large numbers of genes) might raise the level of concern that distinctively human characteristics might be expressed, although most scientists regard this as unlikely.

The already widespread marvel at and openness to human/animal transgenic research underscore the need for Christians to engage this emerging issue now. Although some members of the scientific community propose that transgenic research go forward with few or no restrictions, certain limits are essential. Some will be rooted in the ethical concerns examined earlier. Most basically, however, the complete fusion of animal and human genomes runs counter to the sacredness of human life created in the image of God. However, insertion of a single gene will not transform an animal into a human. Limited transfer, then, can be allowed, although the current trend to insert "more and more" human DNA must be subject to regulation. There is a need to formulate guidelines which will prevent the creation of transgenic animals with uniquely human characteristics. In any case, it appears that natural species barriers still exist and will not be easily overcome, especially those where large deletions and insertions of genetic information are involved.[17]

17. Efficiency rates for creating stable transgenic mice can be as high as 30%, but many other studies report insertion rates of 1 transgenic animal in 1,000 or 1,000,000

A reemphasis on God's created order is needed today, particularly within the Christian community. Furthermore, the Church should encourage young Christians to accept God's calling on their lives for scientific careers in areas such as genetic engineering. Devoted Christians can become outstanding, highly-trained scientists without compromising any biblical values. God's Word is relevant for today with direction for every problem humanity will ever face.[18] It reminds us that the amount of human benefit potentially to be gained is irrelevant if the means run counter to God's divine order. Good ends never justify unethical means. The temptation to create human/animal hybrids in the name of human well-being, therapy, or for knowledge's sake alone cannot justify animal/human hybrids with substantial mixing of genetic material. Much wisdom is needed in the transgenics arena, and it is essential that Christians be actively involved.

---

more embryos. Many deletions and insertions of genetic material result in spontaneous abortion during the embryonic or early fetal development or become lethal soon after birth.

18. In Jeremiah 2:13 God brings two charges against His people, "They have forsaken Me, the fountain of living waters, and they have hewn themselves cisterns — broken cisterns that can hold no water." No matter how sophisticated medical science becomes, the world will still need God as its healer. The Church must encourage people to trust in Him first, then use medical science within godly ethical parameters as a God-given means to bring about healing.

Artificial Intelligence and Personhood

ROBERT K. GARCIA

Recently, a spate of books has been published predicting the rise of robots and intelligent systems that will mimic, if not surpass, the intelligence and independent sentience of human beings.[1] The authors of these works anticipate a time in the not too distant future when humanity's nonbiological progeny replace human beings as the dominant species on Earth, and they speculate on whether people will be forced to merge with their machines to survive. In fact, the concept of a truly artificial mind is one that has captured the fancy of philosophers, computer scientists, and science fiction writers for years. If such a thing could be created, however, the ethical and social implications would be staggering. Would this device or program have a moral claim to "life" and freedom? What would be the responsibility of its creators to the creation? These issues require us to carefully consider what it means to be a person.

What *is* a person? Personhood is one of the most fundamental concepts in ethics. Currently, this concept is under considerable critical review. Writing in the journal *Ethics & Medicine,* Jim Leffel provides an instructive discussion of the three competing definitions of "humanity" implicit in theism, enlightenment rationalism, and postmodernism.[2] Theism defines human personhood essentially: there is a universal human essence, which is in the image of God. Enlightenment rationalism defines personhood rationally: humans are essentially and uniquely rational beings. Postmodernism defines personhood

1. For example, Raymond Kurzweil's *The Age of Spiritual Machines;* Hans Morovec's *Robot: Mere Machine to Transcendent Mind;* Faith D'Alusio, Charles Mann and Peter Menzel's *Robo Sapiens: Evolution of a New Species.*

2. Jim Leffel, "Engineering Life: Defining 'Humanity' in a Postmodern Age," *Ethics & Medicine* 13:3 (1997): 67-71.

socially: humans are distinguished by their social relationships. On this latter view, it is not that humans have a capacity for social relationships; rather, persons are constituted by their position in society and their relations therein — there are no universals or essences, much less a universal human essence.

A fourth view should be added, one prevalent in naturalistic cognitive science, namely, the view that persons are essentially *functional* things. This is not the belief that human persons have functionality. Rather, this is the claim that something's being a person is a matter of its having the right functional relationships — irrespective of its intrinsic properties or its natural kind. Being a person, therefore, is not a matter of having certain *intrinsic* properties or being a certain *kind* of thing. Functionality, furthermore, is specified in terms of causal relationships.

What all this means, more concisely, will be clarified in the pages that follow. This chapter will argue that artificial intelligence (AI) has serious ethical implications, by virtue of its implying a functional criterion for personhood. In particular, AI views are untenable because they require a theory of the mind whose implications conflict with certain ethically-important beliefs which seem to be true or highly justified: (1) the beliefs that certain comparatively immature individuals (e.g., infants) are persons, and (2) that reason and meaning play a significant role in human mental processes (e.g., decision making). Moreover, in opposition to the fundamental premise of AI, it seems impossible that something can be a mind or mental state if it is a computer or computational process.

This chapter is not suggesting that work in cognitive science, AI, or related fields is completely without profit. There is, for instance, exciting and useful work being done using computers to model or study the organic processes of brains and other natural systems. This type of work has been called "Weak AI" by John Searle[3] and does not make a metaphysical claim about whether minds and computer programs are the same kind of thing (it is this latter claim which is at issue here). In this chapter, the term "AI" intentionally excludes Weak AI from its meaning.

AI in the History of the Philosophy of Mind

The issue of AI is important for ethics precisely because, as a theory of the mind, it has implications for personhood. Some historical perspective will be

---

3. John Searle, *The Rediscovery of the Mind* (Cambridge, MA: MIT Press, 1992), pp. 201-2.

helpful here. We may begin with René Descartes, who espoused the concept of substance dualism. In his view, the mind and the body are two distinct substances — one mental and one physical. In recent years, however, substance dualism has been made to wear the sociological dunce cap. Among the many challenges dualism faces, it is often rejected because of an *a priori* commitment to philosophical naturalism, the view that the natural world is all there is. Other people find substance dualism guilty of placing the mind entirely in the private and subjective arena, directly accessible only by its solitary ego, and out of reach of other minds.

In response to this problem, a general view called behaviorism launched a radical overcorrection by placing the mind entirely into the public arena.[4] According to *logical* behaviorism, any statement about the mind is equivalent in meaning to a set of statements about behavior. Thus, for example, the statement "Billy believes that it will rain" is literally synonymous with "Billy behaves as if it will rain." Similarly, *ontological* behaviorism claims that psychological facts are identical to behavioral facts; a pain, for example, is just wincing and groaning. In effect, both versions of behaviorism ignore the internal aspect of the mind. The devastating objection to both versions of behaviorism is that behavior is neither necessary nor sufficient for belief. One can believe something without behaving as if one believes it; and one can behave as if one believes it without actually so believing. The truth conditions for behavioral statements and mental statements are different, and this implies that the statements cannot be synonymous. Thus, to put the point playfully, since behaving as if you believe in behaviorism is neither necessary nor sufficient for actually believing in it, and since believing in behaviorism is neither necessary nor sufficient for behaving as if one believes it, most philosophers have neither behaved as if nor believed that behaviorism is true.

Following behaviorism, historically, came type-identity physicalism, according to which every type of mental state is identical with a type of brain state. Neuronal processes do not *cause* conscious processes — they *are* conscious processes.[5] On this view, a pain is literally identical with a specific state of the brain and/or central nervous system, such as a certain set of C-fibers firing. Although there are several important objections to this theory, one in particular has spawned a new understanding of what is "mental" in psychology and cognitive science.[6] This objection, called the "multiple realization ar-

4. Jaegwon Kim, *Philosophy of Mind* (Boulder, CO: Westview Press, 1998), p. 27.

5. Colin McGinn, *The Mysterious Flame: Conscious Minds in a Material World* (New York: Basic Books, 1999), p. 18.

6. Kim, *Philosophy of Mind*, p. 73.

gument," is based on the intuition that any given mental state can be realized by or in a large variety of physical or biological structures. Thus, type-identity physicalism is guilty of "neuronal chauvinism" by claiming that "unless an organism has C-fibers or a brain of appropriate biological structure, it cannot have pain."[7] This is chauvinistic both against other actual organic systems — whose brains are very different from human brains, and against other possible nonorganic systems.

The question raised by the multiple-realizability of mental states is the following: "What do all pains — in humans, canines, octopuses, and Martians — have in common in virtue of which they all fall under a single psychological category called 'pain'?"[8] What, in other words, *individuates* mental categories? One important and highly influential answer to this question is called functionalism, according to which a mental category is a functional category, and mental states are defined in terms of the causal relationships between a system's external input, causal output, and internal causal relations. The concept of a table, for example, is a functional concept: what makes something a table is its functioning in a certain way, rather than its having a certain physicochemical structure, whether plastic, metal, or wood. Similarly, the concept of pain is a functional concept; Martians, humans, and dogs can all be in pain in virtue of their mental states instantiating the causal role distinctive for pain. According to functionalism, in fact, any system whatsoever, no matter what it is made of, can have mental states provided only that it have the right causal relations between its inputs, its inner functioning, and its outputs. Thus, rather than being characterized by its intrinsic features, each mental state is characterized by the inputs and outputs that constitute its role in a system.[9]

Now, what functionalism has needed is an account of "what it is about the different physical states that gives different material phenomena the *same* causal relations."[10] *How*, in other words, are different physical structures causally equivalent? To this question, the developing science of artificial intelligence offers a powerful and provocative answer: Different material structures can be mentally equivalent if they are different hardware implementations of the same computer program. Thus, neuronal chauvinism is circumvented by claiming that the brain is just one of the many possible computer hardwares that can have a mind. Just as programs are capable of being

---

7. Kim, *Philosophy of Mind*, p. 69.

8. Kim, *Philosophy of Mind*, p. 76.

9. J. P. Moreland and Scott Rae, *Body & Soul: Human Nature and the Crisis in Ethics* (Downers Grove, IL: InterVarsity, 2000), p. 25.

10. Searle, *Rediscovery of the Mind*, p. 43, emphasis mine.

implemented in a variety of intrinsically different hardware, so minds and mental states are capable of being realized in multiple physical systems and states. In this way, functionalism provided the fundamental context for the premises of AI.

Premises and Versions of AI

Before distinguishing two versions of AI, we should note that essential to AI as a whole is the claim that the internal functioning of the mind is computational. Mental states and processes are computational states and processes. That is, the brain is a computer, a system or device whose function is to manipulate symbols according to a program. A program is an algorithm, a systematic procedure for solving a problem in a finite number of steps. So a computer program is a symbol-manipulating algorithm.[11] A brain, therefore, is a computer whose function is to implement programs; and a mental state is an actual function, the implementation of a program.

At this point, it will be useful to employ Searle's distinction between Strong AI and Cognitivism. Whereas Strong AI claims that a mind is *just* a computational state, Cognitivism claims that a mind is *at least* a computational state. Thus, in Strong AI, *all there is* to having a mind is implementing the right program; mental processes are constituted by computational processes. As Searle explains, "any physical system whatever that had the right program with the right inputs and outputs would have a mind in exactly the same sense that you and I have minds."[12] Thus, the slogan: the mind is to the brain as the computer program is to the computer hardware.

Cognitivism, on the other hand, is the view that mental processes, while not wholly constituted by computation, nevertheless have a computational structure. The brain is a digital computer and the mind is, at least in part, a computer program. Thus, while there is *more* to being a mind than just implementing the right program, all minds — whatever else may be true about them — are necessarily computational. An entity cannot be a mind if it is not computational.

These views can be summarized and then critiqued in terms of necessary and sufficient conditions.

11. McGinn, *Mysterious Flame*, p. 178.
12. John Searle, *Minds, Brains, and Science* (Cambridge, MA: Harvard University Press, 1984), p. 28.

Strong AI asserts that computational states are *necessary and sufficient* for minds.

Cognitivism asserts that computational states are *necessary* for minds.

The burden of this paper is to level a two-pronged critique against both Strong AI and Cognitivism. The first prong concerns the ethical implications of each view and the second prong concerns their philosophical shortcomings.

It is important to recognize that the rejection of AI is not a partisan position held only by theists or other non-naturalists. In fact, some of the strongest opponents of AI are outspoken atheistic naturalists (e.g., Searle and McGinn). Whether or not a naturalist can consistently reject AI is an interesting question outside the purview of this paper.[13]

Ethical Implications

The ethical implications of Strong AI and Cognitivism are problematic in that they contradict or undermine commonly held basic beliefs or concepts. This conflict means either that these two views are simply untrue, or that they can only be adopted at great cost, i.e., we must abandon these basic beliefs or concepts. At the very least, the ethical implications indicate the high stakes involved in these issues and a reason for a careful and sober estimation of AI's philosophical warrant.

One way to identify particularly important ethical implications in this case is to consider how the two views construe personhood. The concept of personhood is central to ethics. It is persons who we believe are morally responsible, have certain rights, and enjoy intrinsic value. The importance of personhood is especially evident, for example, in the abortion controversy, where debate often centers around the personhood of the fetus.

Just as personhood is fundamental to ethics, so the mind is fundamental to personhood. Among the things that are taken to constitute personhood are the capacities[14] for intelligence, understanding, sentience, and responsible agency. Since these are *mental* capacities, the nature of the mind will ultimately determine the nature of these capacities.

---

13. Kim, for example, has questioned Searle's consistency on this very point. See Kim, *Philosophy of Mind,* pp. 99-101.

14. I mean here to denote the *ultimate* capacities that are taken to constitute personhood. For a helpful discussion of this subject, see Moreland and Rae, *Body & Soul,* pp. 71-73.

AI Implies a Functional Criterion for Personhood

Since both Strong AI and Cognitivism make being in a computational state a necessary condition for being a mind, the discussion of ethical implications will focus on this shared necessary condition. Later, in the philosophical critique, Strong AI's claim that being in a computational state is a *sufficient* condition for being a mind will be addressed. The fundamental ethical problem with both views is that each implies a *functional* criterion for personhood, according to which the essential properties and/or value of an entity are *not* sufficient for personhood.

For AI, the functional criterion is spelled out in terms of "implementing the right program." Such implementation, however, has nothing to do with the *intrinsic* properties or capacities of the system or entity in question. This is because the intent of functionalism is to account for the multiple forms that mental states can take in a way that avoids neuronal chauvinism against possible minds whose basal structure is intrinsically different from that of human brains. Functionalism stipulates that a mental state is what it is, wholly by virtue of its complex causal relations, and not by virtue of whatever stands in those causal relations. Thus, AI must find a *functionalist* criterion for what it means to "implement the right program."

It is not sufficient that something have the *capacity* to implement the right program. If this capacity alone were sufficient, then, assuming we had the "right" program loaded, we would never have *to turn on* a computer for it to count as having mental states. Rather, it is the actual causal functioning of a computer — the program implementation — that is required.

Thus, *implementing a program* means functioning in such a way that the causal structure and processes can be interpreted as manipulating syntax according to an algorithm. The *right program* presumably means a program whose performance when implemented is indistinguishable from the performance of humans who are intelligent, sentient agents. In other words, the "right program" would have to pass at least the famous Turing Test. The basic idea behind the Turing Test is that if we are to ascribe intelligence and sentience to a computer, it must behave in a way that is indistinguishable from an intelligent human being.[15]

If being a person requires having a mind, and if, according to Strong AI and Cognitivism, having a mind requires implementing the right program,

---

15. For an interesting description of how actual Turing Tests have been performed, see Paul Churchland's *The Engine of Reason, The Seat of the Soul* (Cambridge, MA: MIT Press, 1996), pp. 227-234.

then whatever does not do so is not a person. Herein lies the rub for ethics. AI's functionalist criterion for personhood denies personhood to those who are not functioning "rightly." If implementing the right program is necessary for personhood, and if the "right program" means one whose causal functioning is indistinguishable from the intelligent performance of human persons, then any organism that is not performing up to this standard is not a person.

However, it is obvious that many beings often described as persons do not pass such a test, that is, many are not causally functioning in such a way as to simulate intelligence. Fetuses, prelingual children, adults with localized aphasia, comatose patients, even those who are asleep — none of these beings' causal functions simulates intelligence.[16] Thus, either we must deny that these beings are persons or deny that implementing the right program is necessary for having a mind and being a person. Furthermore, since in Strong AI and Cognitivism the intrinsic properties of a thing are irrelevant in determining whether or not it is a mind, these views make the condition of being essentially made in the image of God *insufficient and irrelevant* for personhood and/or the intrinsic value and rights that attend it. This conclusion would appear to contradict the significance attached to the image of God in the Bible.

AI Implies that Reason and Meaning Are Irrelevant in Mental Processes

Both Strong AI and Cognitivism, by stipulating a necessary condition for minds and mental states, imply that every mental state and process is essentially computational. AI implies that rationality, understanding, sentience, and agency are essentially computational processes. The problem, however, is that this view makes reason and meaning irrelevant to mental processes.

If mental processes are essentially computational, then their constitutive states follow each other not in virtue of their contents standing in certain logical or semantic relationships, but by causal necessity. Thus, a mind does not make logical inferences and does not relate ideas in virtue of their meaning. Rather, a mind is in whatever state it is entirely because of whatever causal laws govern its physical computational processes. Consider a mind that goes through the following thought process. First the mind thinks that:

(1) All humans are mortal, and Socrates is human.

16. Churchland, *Engine of Reason,* p. 234.

After considering (1), the mind thinks that:

(2) Therefore, Socrates is mortal.

We intuit that (2) stands in a logical relationship (that of entailment) with (1). However, if the (1)-to-(2) mental process is computational, then (2) is not a mental state realized in virtue of the logical relationship between (1) and (2). Rather, (2) would be realized entirely in virtue of the fact that the physical properties of the syntax representing (1) nomologically cause the realization of the syntax representing (2).

Thus, if minds are essentially computational, the relationship between (1) and (2) would be causal:

(1) causes (2)

and not inferential:

(1) *implies* (2).

So it is not surprising to hear Jaegwon Kim say that for any computational process, the functional or computational relations among the various abstract parameters — such as the symbols, states, or scanner-printer — can be replaced with appropriate causal relations among the physical embodiments of these parameters.[17]

The same point can be made for the semantic relations between mental states. In a computational process, states do not relate to each other in virtue of whatever meaning might be assigned to them, but in virtue of their syntax. And syntax, of course, is just the shape or properties of a physical state. As Kim observes, "Computational processes respond only to the shapes of symbols; their meanings, or what they represent, are computationally irrelevant."[18]

Finally, a possible representation of an implication is not itself an implication.[19] The fact that (1)-causing-(2) might be interpreted as a logical implication does not make it an implication. After all, the computational relationship between state (1) and state (2) is what it is independent of any interpretation we might assign to it. Thus, even if there could be a rational or semantic insight in the mind over and above the computational process, the insight itself contributes nothing to state (2) coming to be. That is, it would

17. Kim, *Philosophy of Mind*, p. 86.
18. Kim, *Philosophy of Mind*, p. 100.
19. I wish to thank Greg TenElshof for this insight.

be entirely epiphenomenal with respect to the relationships between the computational states governed by the program.

This implication seems to undermine the justification of any belief — whether a belief about what is right and wrong, or a belief about whether the mind is computational. We would not believe things because of or on the basis of reasons; we would believe things because of the causal laws governing the relationships between the physical properties of our mental "syntax." Thus, we must either deny that mental processes are solely computational, or we must give up the commonsense belief that reasons and meaning really do play a role in mental processes.

Philosophical Shortcomings

The foregoing ethical problems are significant. At the very least they render AI suspicious. However, the greatest shortcomings of AI views are philosophical in nature.

Implementing a Program Is Not Sufficient for Being a Mind

No amount of program implementation will constitute or generate a mental state. The reason for this has been put very simply by Searle: syntax is not sufficient for semantics,[20] and symbol manipulation is not sufficient for semantic understanding. A program has no understanding of the symbols it manipulates. This point has been set forth quite poignantly by Searle, in his famous "Chinese Room Argument." The purpose of Searle's thought experiment is to show that a program, qua symbol manipulator, has no understanding of its symbols:

> Imagine that a bunch of computer programmers have written a program that will enable a computer to simulate the understanding of Chinese. So, for example, if the computer is given a question in Chinese, it will match the question against its memory, or database, and produce appropriate answers to the questions in Chinese. Suppose for the sake of argument that the computer's answers are as good as those of a native Chinese speaker. Now then, does the computer, on the basis of this, understand Chinese, does it literally understand Chinese, in the way that Chinese speakers un-

---

20. Searle, *Rediscovery of the Mind,* p. 200.

derstand Chinese? Well, imagine that you are locked in a room, and in this room are several baskets full of Chinese symbols. Imagine that you (like me) do not understand a word of Chinese, but that you are given a rulebook in English for manipulating these Chinese symbols. The rules specify the manipulations of the symbols purely formally, in terms of their syntax, not their semantics. So the rule might say: "Take a squiggle-squiggle sign out of basket number one and put it next to a squoggle-squoggle sign from basket number two." Now suppose that some other Chinese symbols are passed into the room, and that you are given further rules for passing back Chinese symbols out of the room. Suppose that unknown to you the symbols passed into the room are called "questions" by the people outside the room, and the symbols you pass back out of the room are called "answers to the questions." Suppose, furthermore, that the programmers are so good at designing the questions and that you are so good at manipulating the symbols, that very soon your answers are indistinguishable from those of a native Chinese speaker. There you are locked in your room shuffling your Chinese symbols and passing out Chinese symbols in response to incoming Chinese symbols. On the basis of [this] situation . . . there is no way you could learn any Chinese simply by manipulating these formal symbols.

Now the point of the story is simply this: by virtue of implementing a formal computer program from the point of view of an outside observer, you behave exactly as if you understood Chinese, but all the same you don't understand a word of Chinese.[21]

Put simply, *syntax is not sufficient for semantics.*

Remember, a computer program is a symbol-manipulating algorithm. Computers function entirely by virtue of the syntax or physical properties of their states, but do not function by virtue of any meaning or semantic value that may be assigned (by us) to the symbols being manipulated. As Colin McGinn avers, "Programs are semantically blind."[22] *Mental* processes, however, "involve the manipulation of meanings, not merely strings of syntax."[23] As McGinn explains, "Understanding speech is pairing meaning with sounds, not just producing one sound when you hear another one, since that can be done without the assignment of meaning."[24] In other words, minds have understanding, and mental states have intrinsic content or intentionality. Yet, since implementing a program is not sufficient for understanding, and since

21. Searle, *Minds, Brains, and Science,* pp. 32-33.
22. McGinn, *Mysterious Flame,* p. 183.
23. McGinn, *Mysterious Flame,* p. 182.
24. McGinn, *Mysterious Flame,* p. 182.

programs have no intrinsic content, program implementation is not sufficient for minds or mental states.

Implementing the Right Program Is Not Necessary for Being a Mind

A mind or mental state need not be computational. Consider an episode of
sentience, such as a feeling of pain. This is a state of consciousness, a feeling,
but it is not an instance of symbol manipulation.[25] "A feeling is not the same
as a symbol."[26] Besides, *nothing* is necessarily computational. Being a computation is observer-dependent, but it is not an intrinsic fact about a thing. We
assign to the intrinsic properties of something a syntactical interpretation,
and we assign certain meanings to those symbols. We don't *discover* 1s and 0s
in our hardware; we *designate* certain electrical states to *stand* as 1s and 0s.
Thus, nothing is intrinsically computational, since computations are just the
use to which we put intrinsically non-computational states and processes. It
is dubious, therefore, to claim that the mind is essentially computational.

Computational States Are Not Necessary for Mental States Because Thoughts Themselves Cannot be Computational

As previously emphasized, semantic content is extrinsic both to syntax and to
computations that manipulate syntax. Syntax has no necessary semantic assignment; meanings are conventionally assigned to symbols. However,
intentionality is an essential fact about mental states; mental states have intrinsic semantic content. But, if mental states were necessarily computational,
they could not have intrinsic meaning. But they do, therefore they are not
necessarily computational. In sum, since a mental state must have what a
computational state cannot (i.e., essential intentionality), the former cannot
be the latter. Computational states are far from being the kind of thing that
thoughts must necessarily be.

*   *   *

The foregoing points indicate the inadequacy of *any* functionalist theory of
the mind. A functionalist view of the mind undermines critical moral beliefs

25. McGinn, *Mysterious Flame*, 183-4.
26. McGinn, *Mysterious Flame*, 184.

and is philosophically problematic. Though programs may simulate cognitive processes, it is highly doubtful that their creations will truly possess minds or warrant the description as "persons." What requires reconsideration is the viable possibility that the mind is essentially *immaterial,* the seat of our essence, the ground of our value, and the image of God.[27]

27. I am indebted to Nate King and Amy MacLeod for their helpful comments on earlier drafts of this paper.

Cybernetics and Nanotechnology

C. CHRISTOPHER HOOK

In late spring 2000, amidst tremendous fanfare, it was announced that the project to map all the genes of the human cell, the Human Genome Project, was very near completion. Much discussion and literature have been generated concerning the anticipated abilities to heal sickness and potentially improve humanity that this knowledge would provide. Concern has also been raised that this knowledge, and its daughter technologies of genetic screening and manipulation, will be used to select or exclude certain traits or to enhance individuals, possibly even to engineer a new, supposedly superior species of humanity. These issues are extremely important and worthy of the concern and scholarship dedicated to them.

However, the less well-known technologies of cybernetics and nanotechnology are equally sweeping in their potential to transform or recreate human beings into any number and manner of forms. Unfortunately, our conceptions of cybernetics or bionics have generally been formed by Hollywood, and have thus been dismissed as remote fantasies. Most have heard little to nothing of nanotechnology, at least not until the press recently had a field day with Bill Joy's April manifesto of peril and despair. Therefore, the goals of this essay are to review what these technologies are, to explain how far along the path of their development we presently stand, and to explore some of the ethical issues these technologies will present to us.

Cybernetics

The term cybernetics was coined by Norbert Weiner in 1948, and a popularized version of his work was released in 1950 under the provocative title, *The*

52

Human Use of Human Beings.[1] He believed that the key to successful function of both living organisms and machines was the information feedback loop, a process which allows self-regulating activity via continuously updated information about the current status of the machine or organism and its environment. Because both equally depend on this feedback process, Weiner believed machines and living organisms could be blended, combined into a superior device or creature.

In 1960, Manfred Clynes and Nathan Kline took Weiner's concept and considered its use to improve human astronauts for space flight. They coined the term cyborg, or *cyb*ernetic *org*anism, to refer to the blending of technology and humanity. Their speculations resulted in a 1963 NASA report, "Engineering Man for Space: The Cyborg Study."[2]

To the extent that many or most of us depend on technologies such as filled or false teeth, glasses or contact lenses, hearing aids, pacemakers, dialysis, hairpieces, and even vaccinations, we are all cyborgs. But far beyond anything we have seen thus far, we are on the verge of a bold new era of incredible cybernetic enhancements. The following are just a few of the recent developments in the field.

- By the mid-1990s, Peter Fromherz and colleagues at the Max Planck Institute for Biological Cybernetics in Tübingen, Germany had already successfully grown connections between the neurons of several species of animals and transistors, allowing two-way communication through the silicon-neuronal junction.[3]
- In 1999, scientists at MCP Hahnemann School of Medicine and Duke University collaborated to produce a rat with electrodes implanted in its brain, allowing the rat to open a door to get the reward treat merely by thinking about it.[4]
- Also in 1999, researchers at the University of California-Berkeley were

1. See Theodore Roszak, *The Cult of Information: A Neo-Luddite Treatise on High-Tech, Artificial Intelligence, and the True Art of Thinking,* 2nd ed. (Berkeley: University of California Press, 1994), pp. 9-10.

2. Manfred Clynes and Nathan S. Kline, "Cyborgs and Space," *Astronautics* (Sept. 1960): 26-7, 74-5. Also, Robert W. Driscoll, "Engineering Man for Space: The Cyborg Study" (Final Report NASw-512), May 15, 1963.

3. Peter Fromherz and Alfred Stett, "Silicon-Neuron Junction: Capacitive Stimulation of an Individual Neuron on a Silicon Chip," *Physical Review Letters* 75 (1995): 1670-1673. Also, S. Vassanelli and P. Fromherz, "Neurons from Rat Brain Coupled to Transistors," *Applied Physics A.* 65 (1997): 85-88.

4. Victor Chase, "Mind Over Muscles," *Technology Review* 103 (March/April 2000): 44.

able to measure the neuronal activity of 177 cells in one of the optical pathways of a cat, process the information, and re-create rough images of what the cat's eyes were viewing at the moment.[5]

- The new millennium arrived with a report that the Dobelle Institute had used neural implants to enable a sixty-two-year-old man, blind since the age of thirty-six, to perceive rough images composed of small dots of light, allowing him to read large letters and navigate around big objects.[6]
- Meanwhile, investigators at Emory University have helped two patients with locked-in syndrome, a state in which the brain is conscious but cannot produce any movement of the patient's voluntary, skeletal muscles. The unfortunate patient is often thought to be in a persistent vegetative state. The two received brain implants, into which their neurons grew, establishing a link with a computer. This enabled the patients to use their minds to control a cursor on a computer screen and thus communicate with others.[7]
- Richard Hahnloser and colleagues have reported the successful creation of a silicon chip that uses digital selection and analogue amplification in mimicry of the nerve cells of the brain.[8] Ultimately such devices will allow easier integration of cybernetic prostheses with the brain.
- In Chicago in June 2000 the first artificial retinas made from silicon chips were implanted in the eyes of three blind patients suffering from retinitis pigmentosa. Each implant was 2mm in diameter, $\frac{1}{1000}$ of an inch thick, and contained ~3500 microphotodiodes that convert light energy into electrical impulses to stimulate the remaining functional nerve cells of the retina.[9]
- For many years Steve Mann, now at the University of Toronto, has been perfecting "wearable computers." They have now achieved the degree of miniaturization that the devices look like part of a normal wardrobe. The purpose of these devices is to provide constant access to computer networks. He writes, "Every morning I decide how I will see the world

5. Garrett Stanley, Fei F. Li, and Yang Dan, "Reconstruction of Natural Scenes from Ensemble Responses in the Lateral Geniculate Nucleus," *Journal of Neuroscience* 19 (Sept. 1999): 8036-8042.

6. Associated Press, reported in CNN.com, January 16, 2000.

7. Chase, "Mind Over Muscles," pp. 39-45.

8. Richard Hahnloser, et al., "Digital selection and analogue amplification coexist in a cortex-inspired silicon circuit," *Nature* 405 (June 22, 2000): 947-951.

9. The device was developed by Alan and Vincent Chow and is reported on the website for Optobionics Corporation at http://www.optobionics.com/.

that day. Sometimes I give myself eyes in the back of my head. Other days I add a sixth sense, such as the ability to feel objects at a distance. . . . Things appear differently to me than to other people. In addition to having the Internet, massive databases and video at my beck and call most of the time, I am also connected to others. While I am grocery shopping, my wife — who may be at home or in her office — sees exactly what I see and helps me pick out vegetables. She can imprint images onto my retina while she is seeing what I see." He goes on to extol the benefits of always being able to appear the expert on any topic because of his access to vast amounts of information almost instantaneously.[10]

Mann has also recruited others into his new way of living. He led a class of twenty students, teaching them how to use and blend with their "personal imaging and photoquantigraphic image processing" devices, and notes that sixteen of the twenty did not return the "xybernaut" computers at the end of the course. He states, "All have been marked for life. They have been assimilated," a chilling reference to a race of cybernetic beings that inhabit the Star Trek universe of the twenty-fourth century. The Borg, a race of cybernetic individuals characterized on Star Trek, expand their race and culture by "assimilating" other races and individuals, reducing them to drones, which then become part of the larger one-minded collective, their individuality destroyed and their knowledge absorbed to service the whole.

It is not cybernetic implants, however, that initiate the assimilation process of the Borg; these are added later to improve connectivity with the collective and provide various prostheses to assist each drone in the performance of specific duties. Rather, assimilation begins from within, as the body is remodeled, literally cell by cell and sometimes molecule by molecule by microscopic nanorobots that are injected into the victim's body. This is, of course, science fiction for the moment, but nanotechnology, upon which this concept is based, is a real and growing field.

Nanotechnology

The term nanotechnology was coined in 1974 by Japanese researcher Norio Taniguchi to mean precision machining with tolerances of a micrometer or

---

10. Steve Mann, "Cyborg Seeks Community," *Technology Review* 102 (May/June 1999): 36-42.

less.[11] It was Eric Drexler, however, who brought the word and the concept into public consciousness with his 1986 book, *Engines of Creation*.[12] This was followed by a scholarly, technical feasibility study entitled "Nanosystems."[13] To further his dreams, Drexler created the Foresight Institute[14] dedicated to the development of nanotechnology, as well as other life extension technologies.

But the real originator of the idea of nanotechnology, which deals literally with engineering on the molecular and atomic level, was physicist Richard Feynman. He presented his ideas in a speech entitled "There's Plenty of Room at the Bottom," an address given on December 29, 1959 at the annual meeting of the American Physical Society. Dr. Feynman spoke of printing the entire twenty-four volumes of the Encyclopedia Britannica on the head of a pin, such that it could be read by an electron microscope. The scale required to accomplish this feat is literally on the nanometer scale, that is one billionth of a meter. As a point of reference, DNA is 2.3 nanometers wide. Ten shoulder-to-shoulder hydrogen atoms span 1 nanometer. Yet Feynman stated, "I am not inventing anti-gravity, which is possible someday only if the laws are not what we think. I am telling you what could be done if the laws are what we think; we are not doing it simply because we haven't gotten around to it."[15] So what has been achieved in the quest for molecular engineering thus far?

In 1999, two groups of researchers independently fabricated a transistor out of a single carbon molecule.[16] Another group has made a molecule that rotates, acting as a nanowheel, as well as a rudimentary abacus with single molecules acting as the sliding beads.[17] Two groups, reporting in the same Sept. 9, 1999 issue of *Nature*, have described building molecular motors.[18]

11. David Voss, "Moses of the Nanoworld," *Technology Review* 102 (March/April 1999): 61.

12. Eric Drexler, *The Engines of Creation* (New York: Anchor Books, 1986). The entire text of the book may be found online at http://www.foresight.org/.

13. Eric Drexler, *Nanosystems: Molecular Machinery, Manufacturing, and Computation* (New York: John Wiley & Sons, Inc., 1992).

14. The Foresight Institute web page is http://www.foresight.org/.

15. Richard P. Feynman, "There's Plenty of Room at the Bottom." The text of the talk is available online at http://www.zyvex.com/nanotech/feynman.html.

16. David Rotman, "Will the Real Nanotech Please Stand Up?" *Technology Review* 102 (March/April 1999): 46.

17. Rotman, "Nanotech," 46.

18. T. Ross Kelly, Harshami De Silva, and Richard Silva, "Unidirectional rotary motion in a molecular system," *Nature* 401 (September 9, 1999): 150-152; and Nagatoshi Koumura, et al., "Light-driven monodirectional molecular rotor," *Nature* 401 (September 9, 1999): 152-157.

One method of pursuing the development of nanomolecular devices is to use molecular tools already found in nature in new ways. Montemagno and colleagues have successfully tested the use of an F1-ATPase motor protein as a source of power for nanomolecular devices.[19] ATP is one of the most plentiful molecules in cells and critical to cellular energy storage and utilization.

Edwin Jager and colleagues have developed a synthesized microrobot that could move micrometer-sized objects about in a directed way in a liquid medium. Though not a true nanomolecular device, these robots could manipulate single cells or cell-sized particles about in an area of 250x100 micrometers.[20]

An important part of any nanorobot will be control and information processing systems, creating the necessity for nanoscale or molecular computing. In 1994, Leonard Adleman of UCLA reported in *Science* the use of DNA to encode a small graph in order to solve a combinatorial mathematics problem known as the Hamiltonian path problem.[21]

The February 2000 *Proceedings of the National Academy of Sciences* contained an article by Laura Landweber of Princeton describing the use of an RNA computer to solve a chess problem: list all the possible arrangements of any number of knights on a chessboard so that no knight is threatening another. The board was restricted to 9 squares to keep the possible knight arrangements to only 512.[22]

Aside from solving chess problems microscopically, what would nanotechnology really do for us? Although a multitude of manufacturing and engineering uses have been proposed, we will focus on the potential medical uses. Primarily, nanorobots would serve as implantable devices that could:

- Detect and destroy malignant cells and cancers;
- Detect and combat infection as immune machines;
- Detect and repair genetic mutation or injury;
- Deliver targeted drug therapy via synthesis and administration within the body;
- Replace cellular structures with stronger or more efficient materials;

19. Carlo Montemagno, et al., "Constructing Biological Motor Powered Nanomechanical Devices," *Nanotechnology* 10 (1999): 225-231.

20. Edwin Jager, Olle Inganas, and Ingemar Lundstrom, "Microrobots for Micrometer-Size Objects in Aqueous Media: Potential Tools for Single-Cell Manipulation," *Science* 288 (June 30, 2000): 2335-38.

21. Leonard M. Adleman, "Molecular Computation of Solutions to Combinatorial Problems," *Science* 266 (November 11, 1994): 1021-24.

22. Dirk Faulkamen, Anthony Catres, Richard Lipton, and Laura Landweber, *Proceedings of the National Academy of Sciences* 97 (February, 2000): 1385-1389.

- Repair or replace damaged tissues and non-cellular connective tissue materials such as the extracellular matrix;
- Replace or augment physiologic functions;
- Remove atherosclerotic plaque in coronary and cerebral arteries;

Tools like these could significantly prolong the lives of many people, reduce much disability and suffering, enhance quality of life, and even reduce certain health care costs. In fact, many of the nanovisionaries see nanotechnology as an important tool to forestall aging, or in the most questionable predictions, to achieve immortality.

Robert Freitas, a pioneer in the field of nanomedicine, has already published his rudimentary designs for "respirocytes," artificial red blood cells that can deliver oxygen to tissues, as a substitute for regular erythrocytes, or red blood cells.[23] He has gone on to develop plans for a "clottocyte," or artificial platelet. While these designs await the technological developments necessary to create the required molecular components, Freitas has been vigorously thorough in his evaluation of the physical, biochemical, and engineering issues and limitations for such projects, and has found them feasible, just as Feynman predicted. In 1999, he released an immense 509-page treatise, the first of three planned volumes on the feasibility and implementation of nanomedical devices.[24]

Freitas' work is but one example of the serious commitment to nanotechnology already taking place. In 1997 the Department of Defense Military Health Services System published a large study on the use of nanotechnology in medicine.[25] The report also considered what would be required to counter threats from nanodevices that could be developed as biological weapons. In 2000, President Clinton announced the National Nanotechnology Initiative, devoting $270 million in 2000 and $497 million in 2001 to nanotechnology research.[26] At least twenty-three major universities have made investments in nanotechnology research in the last few years. Many of the giants of industry have initiated nanotechnology research programs including IBM,

23. Robert A. Freitas, "Respirocytes: Exploratory Design in Medical Nanotechnology: A Mechanical Artificial Red Cell," *Artificial Cells, Blood Substitutes and Immobil. Biotechnology* 26 (1998): 411-430.

24. Robert A. Freitas, *Nanomedicine: Volume 1: Basic Capabilities* (Austin: Landes Bioscience, 1999).

25. Department of Defense, Military Health Services System, *2020 Study of Biotechnology and Nanotechnology,* 1997. (This is accessible online via the Foresight Institute web site.)

26. National Nanotechnology Initiative, http://www.nano.gov/.

Hewlett Packard, Texas Instruments, and Xerox. Three new technical journals devoted to the field were launched in 1999. The elder statesman, *Nanotechnology*, published by the Institute of Physics, however, is celebrating its tenth anniversary. Without question, this is a viable and exploding field.

Ethical Questions and Issues

Both cybernetics and nanotechnology are extremely exciting areas of study. They promise phenomenal medical therapies. With cybernetics we have the opportunity to help the blind to see, the lame to walk, and those bound and limited by physical or neurological afflictions to interact more dynamically with their world, even to overcome their limitations. Nanotechnology may help us fight cancer, vascular disease, and even infectious disease.

But both technologies have their potential dark sides. It is one thing to use technologies to repair an injury or address an affliction, but quite another to use them for engineering "better" human beings. There is clearly going to be an attraction for those who are well to enhance themselves for a competitive edge via cybernetics, or to increase their longevity via nanotechnology.

Therefore, some of the first ethical questions stimulated by these fields are the same as those being raised about genetic engineering. Who will or should have access to these techniques? Can we clearly define a line of demarcation between when a technology is used for treatment purposes and when it is used for enhancement purposes? For example, issues of height, strength, visual acuity, and intelligence are often described as simple attributes. But to some, their allotted "attributes" are seen as disabilities, inducing significant suffering. Even if we could make clear definitions and distinctions, are there sufficient reasons to prevent the use of the technologies for enhancement? Isn't it a common desire, if not a laudable goal, for each of us to improve ourselves and our children as much as possible?

Many of the proponents of nanotechnology envision a world of prosperity and material sufficiency for all. But aren't we more likely to see different classes or groups arise: the enhanced vs. the unenhanced, inferior, naturally-based model; *homo sapiens* vs. "techno" sapiens? Won't we see discrimination against the unenhanced? As Mark Hanson has recently written, "enhancement technologies, on the whole, progressively redefine normality as defect and ultimately devalue the created self."[27]

27. Mark J. Hanson, "Indulging Anxiety: Human Enhancement from a Protestant Perspective," *Christian Bioethics* 5 (1999): 121-138.

This tendency to disdain the less than ideal is illustrated by the comments of Dr. Robert Ruben in his Presidential Address to the American Society of Pediatric Otolaryngology. Ruben drew a link between hearing loss and criminal behavior. He denounced those parents who would not allow their deaf children to have cochlear implants, warning that those with communication disorders present a threat to the progress and prosperity of America because they are "economically burdensome and destructive of the social fabric."[28]

Humanity remains fallen. Sin is very real and its power to corrupt our thoughts, our judgment, and our desires remains deeply prevalent throughout our species. We should expect our technologies to reflect this sin and magnify the consequences of error. While some see technology as the means of leveling the playing field, making all equal, history has shown that technology tends to produce larger disparities between cultures and subcultures. People often try to find ways to distinguish themselves from others and to define those who are "in" and those who are "out." And we must remember as well that attempts thus far at creating a utopia, where there are prosperity and freedom for all, have typically ended in tyranny. As long as sin remains alive among us, this will be the case.

Some see these evolving technologies as a means to free themselves and humanity from the finitudes of our bodies. This line of thinking has become known as trans-humanism or posthumanism, and is the demon child of the marriage of the worst elements of modern and postmodern thought. From modernism we take an almost blind faith in inevitable progress, with the good defined as the suppression, replacement, and/or total control of the "natural" via science and technology. Radical autonomy is the creed, declaring the right and duty of all people to control their own destinies, to engineer their own evolution. From postmodernism and its rejection of objective truth and the notion of a true, identifiable self is taken the belief that there is nothing intrinsically valuable about the biological form, particularly not the human form. Because there are no true norms for existence or behavior, we may create any reality we desire, and change ourselves in any manner to our suiting. There is no image of God to consider since there is no God who has created us to have any particular characteristics or character. We can foster whatever image we can (be tempted to) imagine.

---

28. R. J. Ruben, "Critical Periods, Critical Time: The Centrality of Pediatric Otolaryngology," *Archives of Otolaryngology, Head and Neck Surgery* 122 (1996): 234-36, quoted in Kenneth J. Dormer and Ashton Whaley, "Ethical Issues in Cochlear Implantation of Children," *Ethics & Medicine* 16 (2000): 11-14.

Katherine Hayles has identified four key assumptions behind posthuman thought:

- First, the posthuman view privileges information pattern over material form, so having a biological form is seen as an accident of history rather than an inevitability of life.
- Second, the posthuman view considers consciousness a mere product of biology, not something beyond it.
- Third, the posthuman view thinks of the body as the original prosthesis we all learn to manipulate, so that extending or replacing the body with other prostheses becomes a continuation of a process begun before we were born.
- And fourth, the posthuman view configures the human being so that it can become interchangeable with intelligent machinery. In the posthuman, there are no essential differences or absolute demarcations between bodily existence and computer simulation, cybernetic mechanism and biological organism, robot technology and human goals.[29]

She concludes her book, entitled *How We Became Posthuman,* with the following statement: "Humans can either go gently into that good night, joining the dinosaurs as a species that once ruled the earth but is now obsolete, or hang on for a while longer by becoming machines themselves. In either case . . . the age of the human is drawing to a close."[30]

And she is not alone in her views: Richard Jastrow, in his book, *The Enchanted Loom,* writes of a time when

a bold scientist will be able to tap the contents of his mind and transfer them into the metallic lattices of a computer. Because mind is the essence of being, it can be said that this scientist has entered the computer, and that he now dwells in it. At last the human brain, ensconced in a computer, has been liberated from the weakness of the mortal flesh. . . . It is in control of its own destiny. . . . It seems to me that this must be the mature form of intelligent life in the Universe. Housed in indestructible lattices of silicon, and no longer constrained in the span of its years by the life and cycle of a biological organism, such a kind of life could live forever.[31]

29. N. Katherine Hayles, *How We Became Posthuman: Virtual Bodies in Cybernetics, Literature, and Informatics* (Chicago: University of Chicago Press, 1999), pp. 2-3.

30. Hayles, *How We Became Posthuman,* p. 283.

31. Richard Jastrow, *The Enchanted Loom: Mind in the Universe* (New York: Simon & Schuster, 1984), pp. 166-167.

This is also the thesis of a recent book by Ray Kurzweil entitled *The Age of Spiritual Machines.* Kurzweil, following the implications of Moore's law (the computational speed of computers doubles every 18-24 months), believes we will soon be able to analyze and replicate the entire synaptic structure of a person's brain by digital simulation, allowing the person to be "uploaded" into a machine. He declares, "we will be software, not hardware."[32] The actual feasibility of this happening is far from certain. What is frightening is that there are folks like Kurzweil and Jastrow who actually see this as a desirable goal.

Virginia Postrel, in her book, *The Future and Its Enemies,* states, "the biological doesn't have the corner on 'natural'. If nature is itself a dynamic process rather than a static end, then there is no single form of 'the natural'. An evolving, open-ended nature may impose practical constraints, but it cannot dictate eternal standards. It cannot determine what is good. The distinction between the artificial and the natural must lie not in their source — human or not — but in their characteristics, in the way they relate to the world around them."[33] Consequently, such thinking would claim that the *imago dei,* the image of God given to humanity, even if it did exist, is irrelevant to any discussion of right and wrong, or to issues of moral rights and standing.

Then there is Kevin Warwick writing in *Wired* magazine: "I was born human. But that was an accident of fate — a condition merely of time and place. I believe it's something we have the power to change."[34] Compare this with a line from Nietszche's *Thus Spake Zarathustra,* "I teach you the overman. Man is something that is to be overcome."[35] Indeed, transhumanist thought is nothing more than the Nietzschean ideal of the will to power.

But it is not just the human form that is to be challenged and overcome; it is the very idea of a self that must be destroyed as well. If we can through our cybernetic enhancements enter worlds of virtual reality, the self becomes plastic, malleable into anyone or anything we desire. Jeri Fink writes in *Cyberseduction,* "Virtual reality simulations, as copies of actuality, can become our new reality. We can 'play' with our conflicts, rehearse and reenact them with the fantasy of achieving a resolution. Families can be perfect, enemies can be murdered and we can rise to become . . . the heroes and heroines of our

32. Ray Kurzweil, *The Age of Spiritual Machines: When Computers Exceed Human Intelligence* (New York: Viking, 1999), p. 129.

33. Virginia Postrel, *The Future and Its Enemies* (New York: The Free Press, 1999).

34. Kevin Warwick, "Cyborg 1.0," *Wired* 8.02 (February, 2000): 145.

35. Friedrich Nietzsche, *Thus Spake Zarathustra,* trans. by Walter Kaufmann (New York: The Modern Library, 1995), p. 12.

worlds."[36] Fink then proceeds to document in page after page examples of in-dividuals who already are using the Internet to escape their real lives, creating new personae, changing sexes, engaging in all sorts of activities they would never consider in the "real" world. A new Edenic apple is once again offering a means to rise above our small, limited selves and allow us to be as God.

Frightening as the thought of the loss of self into alternate realities may seem to many, a great number of others unfortunately will (and already do) find this attractive and be drawn right in. One of the great challenges of grow-ing up, living as an adult with mistakes and consequences and living sin and guilt, is the reality of the responsible self. If one can deny the self, one can seemingly escape responsibility, and thus escape guilt as well. The attempt to hide one's self is a pattern of behavior that dates back to those who chose to take the first such apple offered for enhancement purposes. After Adam and Eve sinned, they took plant leaves and made garments for themselves and then hid themselves among the plants and animals of the Garden. Often, the mak-ing of clothing is attributed to a desire to cover the sexual organs. More impor-tant, however, is the fact that these garments were made of leaves, created to hide their selves, to blend in with their surroundings of "other" plants and leaves.[37] A self cannot be judged if the self cannot be found. Yet the reality re-mains that we are each created as a self, a self who can enter into relationship with God's self and with other selves. Ultimately unable to escape the reality of their self even in their cyberworlds, fugitive cybernauts will then try to flee by plunging into deeper and deeper experiences of virtual sensuality to try to soothe the pain of the self separated from its true source and fulfillment, God. One such escape, cybersex, has already become an addiction for many, de-stroying marriages and interfering with jobs, and can only grow more devas-tating with the increased integration of our minds and cybertechnology.

We must also be cognizant of other concerns. Not only will our cyber-netic connectedness provide opportunities for our access to others and infor-mation, but it will also provide opportunities for others to have access to us. How much more will individuals be subject to those who wish to control and influence them? Will we be able to separate out and eliminate images, in-structions, or "thoughts" meant to influence us, both from commercial and governmental sources? How much further will our privacy erode when the last bastion of our privacy, our mind, is open to the cybernetic web? And as a

36. Jeri Fink, *Cyberseduction: Reality in the Age of Psychotechnology* (Amherst: Pro-metheus Books, 1999), p. 156.

37. Brooks Alexander, "The Faustian Bargain: Computers and Human Potential," *SCP Journal* 19 (1995): 46-70.

further danger, will there be new types of electronic viruses that can damage our brains as well as the cybernetic equipment we are "attached" to?

A team at Los Alamos National Laboratory has been creating a new type of web server that continually rebuilds itself to adapt to users' needs. It puts in new hyperlinks whenever it thinks they'll open up a path that surfers are likely to use, and closes down old links that fall into disuse, essentially resembling the connections that grow and fade in the human brain. This system is in essence the first step to creating a "global brain" that could be implemented in as little as five years from the time of this writing. One of the creators of this type of system sees the global brain as the center of what he calls a global superorganism. Human society will become more like an integrated organism, like an ant colony, with the Web playing the role of the brain and the people playing the role of the cells in the body. Each individual user can be quickly identified as to his or her knowledge base or potential contributions to the whole by means of software devices known as "cookies" that imbed themselves in a given individual's system and can be used to track areas of interest, transactions, etc. A simple form of a cookie is now used by online sellers like Amazon that know you when you sign on and give you a list of "things that might be of interest to you." This brain will, of course, have to protect the body and ensure it has what it needs to thrive and grow. As more and more of our business transactions, power and water supply, etc., become computer-controlled, imagine the coercive power a system like this could have on a person's life and individuality.[38]

Nanotechnology raises some of the same issues as cybernetics — issues raised by the quest for immortality and the remaking of humanity. However, it also poses its own unique threat: that of creating devices that could destroy life on a vast scale. Just as nanorobots could be used to repair injured cells, they can also be designed as killers — microscopic mechanical plagues that could be carried by the winds or the water supply. Nanorobots called assemblers would not only perform a specific function, but would be able to replicate themselves swiftly. As nanopioneer Eric Drexler has stated, "dangerous replicators could easily be too tough, small and rapidly spreading to stop. . . . We have enough trouble controlling viruses and fruit flies. . . . This threat has become known as the 'gray goo problem' [and] the gray goo threat makes one thing perfectly clear: We cannot afford certain kinds of accidents with replicating assemblers."[39]

It is this potential for mass destruction of life that led a leader in the

38. Michael Brooks, "Global Brain," *New Scientist* 166 (June 24, 2000): 22-27.
39. Drexler, *Engines of Creation*, ch. 11.

field of new computer software technologies, Bill Joy, to voice his concerns in a major article in the April 2000 issue of *Wired*. Joy is the cofounder and chief scientist of Sun Microsystems. He writes,

> Most dangerously, for the first time, these accidents and abuses are widely within the reach of individuals or small groups. They will not require large facilities or rare raw materials. Knowledge alone will enable the use of them. Thus we have the possibility of not just weapons of mass destruction but knowledge-enabled mass destruction, this destructiveness hugely amplified by the power of self-replication. . . . I think it is no exaggeration to say we are on the cusp of the further perfection of extreme evil, an evil whose possibility spreads well beyond that which weapons of mass destruction bequeathed to the nation-states, on to a surprising and terrible empowerment of extreme individuals.[40]

Joy then proposes that the only way we can prevent the potential devastation is to stop the research now, to choose not to go down the road at all. Once knowledge exists, it will be used by someone. Therefore, stopping now is the only way to protect the knowledge from being available for misuse.

Joy's comments have been severely criticized throughout the press and the Internet, not because his concerns are misplaced, but because his "solution" is naïve (and, though unstated, is against the prevailing Religion of Triumphant Progress, and doubts the beneficence of its deity, Science). Realistically, research is a very hard, if not impossible, thing to restrict, particularly if private individuals provide the funding. And there will be plenty of millionaires seeking a supposed immortality who will be happy to serve as financiers. We need think only of Saddam Hussein to find an individual who would exploit the development of nanotechnology for purposes of destruction, in spite of supposed bans. Meanwhile, with a ban on research in peaceloving environments, appropriate defenses would not be developed. In 1875, Great Britain, then the world's sole superpower, was sufficiently concerned about the dangers of the new technology of high explosives that it passed an act barring all private experimentation in explosives and rocketry. Consequently it was German missiles that bombarded London in World War II.[41]

Recognizing that research will proceed, we should move forward, developing nanotechnology for its therapeutic benefits while concurrently identi-

---

40. Bill Joy, "Why the Future Doesn't Need Us: Robotics, Genetic Engineering, and Nanotech," *Wired* 8.04 (April 2000): 238-247.

41. Glenn Harlan Reynolds, "Techno Worries Miss the Target," at http://www.intellectualcapital.com/issues/issue381/item9612.asp.

fying ways of detecting and inactivating harmful nanodevices. Recommending a similar course, the Foresight Institute has just released their "Guidelines on Molecular Nanotechnology."[42] While their prospective efforts to understand and regulate this new technology are praiseworthy, some of the statements in their guidelines are naïve as well — especially their belief in the beneficence of industry and the adequacy of self-regulation. There is no way to ensure that these or any guidelines will be universally followed. Not everyone who creates nanomachines will employ the proposed internal devices or design restrictions recommended to make the machines limited in their ability to do harm, or easy to inactivate. Rather, active defense mechanisms can ultimately be the only true protection.

Concluding Reflections

Biotechnologies are rapidly accelerating ahead of sufficient ethical reflection and appropriate plans for control. We must commit ourselves wholeheartedly to analyzing the implications of devices yet to come and prepare to deal with and contain them as necessary. We can no longer afford to deal with these concerns reactively, as has been the model of bioethics over the past several decades. We must apply our scholarship prospectively, looking at what is or may be coming in the next five, ten, twenty years and beyond.

With technological threats that can alter the very nature and or even existence of humanity — and the prevailing perspective of a materialistic, postmodern outlook — our world desperately needs the truth of the Gospel. In Matthew 5:48, Jesus says, "Be perfect, as your heavenly Father is perfect." Attempting to better ourselves is not necessarily wrong, but our focus is to be on spiritual perfection, not physical perfection. Jesus later adds in Matthew 6:31, "Be not anxious," indicating that there are limits to the degree to which we should try by our own might to pursue this or any improvement. We cannot achieve God's perfection by our own means, but only by our submission to His. Paul records God's words to him, "My grace is sufficient for you, for my power is made perfect in weakness" (II Corinthians 12:9). He himself later states in Philippians 4:11, "For I have learned in whatever state I am to be content." Faced with the stark reality of our limitations, we can rest in God, finding our worth not in our performance, but in Him.[43]

---

42. Foresight Guidelines on Molecular Nanotechnology: Monterey Workshop Draft 3.6 (May 14, 2000) at http://www.foresight.org/NanoRev/Guidelines.html.
43. Hanson, *Christian Bioethics*, pp. 121-138.

We must pursue God's vision of perfection for us, not our own vision corrupted by sin. God made us selves — separate persons so that we can enter into relationship with Him and each other. Only as selves can we know love. It is so often the case that "What is exalted by people is an abomination in the sight of God" (Luke 16:15). Accordingly, any treatment or enhancement we choose to employ, indeed anything we choose to pursue in this life, must answer a crucial question: Does the choice at hand, either for myself, or for others, aid in the pursuit of the kingdom of heaven? Truly, those who claim the name of Christ must let the vision of II Corinthians 3:18 guide them: "We all . . . beholding the glory of the Lord, are being changed into His likeness." Do we need these technologies to live as Jesus would before and for the world?

In 1776, Thomas Paine wrote in "Common Sense," "We have the power to begin the world all over again." Indeed, if the predictions of the scholars quoted here are correct, and the trends in technological development continue, we have that power right now. We can be grateful for wonderful new technologies that help us pursue the mission of compassion modeled for us by Jesus as He healed others. But we must not forget that He only restored that which was lost by illness and the effects of sin. He did not make people smarter, or stronger, or encourage them to pursue an earthly immortality. Rather He did all these things to help people return to the Father and seek eternal life with Him. We must not let ourselves be deluded into thinking that technology will solve what at its heart is the spiritual problem of the fall and separation from God. We are being offered a Faustian bargain to create our own worlds in our selfish and sinful image, and we must reject it.

Today, perhaps more than at any time in history, Christians must live authentically as Christians. We must show not a fear of technology, but a courageous control of technology, and refuse to let technology control us. We may enjoy these tools' legitimate fruits, using them for the good of all. But we must at every turn reject the virtual and its influence on us, pursuing instead the authentic, reflecting the joy and fulfillment of real relationships, real life, and real freedom in Jesus Christ. There is a real alternative to the anxiety of our culture, the exhausting barrenness of unending competition, the emptiness of loveless sensualism, real or virtual, and the pursuit of false immortality. It is the living, breathing body of Christ. Christians must commit themselves to holy transformation so that others who look at them will see not virtual love, or virtual Christianity, but the real thing, and then desire to be a part of it.

In *The Abolition of Man*, C. S. Lewis wrote, "There neither is nor can be any simple increase of power on Man's side. Each new power won *by* man is a power *over* man as well. Each advance leaves him weaker as well as stronger.

In every victory, besides being the general who triumphs, he is also the prisoner who follows the triumphal car."[44]

Let us choose faithfully to find our triumph in the sufficiency and mercy of our Lord Jesus Christ. And let us pray, as did King Arthur,

> May God give us
> > The Wisdom to discover the Right,
> > The Will to choose it, and
> > The Strength to make it endure.

44. C. S. Lewis, *The Abolition of Man* (New York: Simon & Schuster, 1947), p. 69.

PART II

GROWING CULTURAL
CHALLENGES

Multiculturalism

JOHN PATRICK

There is an almost total and unthinking acceptance of multiculturalism by students and many faculty members on college campuses today. It is accepted as a given of modern society. At one level this is appropriate. Ethnic diversity is increasing. However, it is important to emphasize that this fact does not necessarily translate into cultural diversity in ways that are important for public policy or professional ethics. For that to happen there has to be a much higher correlation between ethnic origin and belief system.

The 1991 Statistics Canada Census asked about belief systems; a list of all the known belief systems was provided and Canadians were asked to indicate where they fit. The results suggest that they may have used a process of elimination ("I am not a Zoroastrian, not a Jew, not a Muslim, certainly not a nothing ['no belief'], not an atheist; then, O dear, I must be a Christian"). Despite increasing immigration, insofar as Canadians could identify their belief system, it turned out to be ninety percent Christian. Public policies in this democratic nation do not reflect such a Christian commitment. Not all minorities have different beliefs; many still inhabit the Christian explanatory story and consequently have little problem with negotiating the cultural milieu. Immigrants from other cultures often migrate for reasons that make them highly likely to adopt the culture into which they are immigrating. For example, a significant number of the Asian immigrants to Canada are highly committed Christians rather than Buddhists.

The assumptions underlying multiculturalism are not well established and its claim to be the only way to allow disparate groups to live together peacefully without imposing any one view is manifestly untrue. The rules are formulated by an elite group, who, by and large, do not believe in God and wish him to be removed from the public square. The rules therefore impose, and lead inevitably and unsurprisingly to, the result desired. In its more radical versions multiculturalism is culturally homogenizing and can be particularly inappropriate for a profession such as medicine, which has traditionally been understood as a moral activity and deals with people as individuals and not as groups. Patients are always people who belong to particular cultures; no one has ever seen nor will ever see a multicultural person. The optimal practice of medicine requires that the physician understand the strengths and weaknesses of the patient's cultural background in order to use the strengths and avoid the weaknesses in the care of the patient. It is certainly insensitive to treat humanists as though they are Christians, but it is equally insensitive to assume, as multiculturalism does, that a bland form of humanistic belief will serve for devout believers. Multiculturalism is intrinsically syncretistic and homogenizing but patients are particular and it is important to consider the details of their cultures in many medical settings. Even within the Christian community it is important to understand the subculture. Resistant hypertension can be the result of naive views on the nature of spiritual healing rather than merely a biological problem!

Cultures: Equal or Different?

Despite these obvious realities, most students and academics think that multiculturalism is a good basis for public policy. They are invariably surprised to find another academic who not only disagrees with their position but is also not at all unwilling to deconstruct that position. The first question in this process of deconstruction is: "Are all cultures equal?" The temptation, which flows from some versions of multiculturalism, is to say "yes," for to say "no" may lead to a label of ethnocentricity or worse, Eurocentricity. Nevertheless there is a vague uneasiness that something is amiss. Further conversation reveals that this is only a felt, not a cognitive phenomenon; most people cannot (or perhaps will not) articulate it often for fear of the consequences.

In these days of political correctness, questions and stories are the safest approaches to engaging these sensitive issues. I invariably tell the same true story from my own medical practice. It goes to the heart of the matter immediately. A few years ago I saw a child with a septic knee. The knee had been totally destroyed because the arthritis had been neglected. By the time the child arrived at the hospital the only option left was amputation. The parents were told this and requested time to think about it. They were told that they must not think for long as the child was septic and would soon die if nothing were done. They returned within half an hour or so having decided that they would not allow the amputation and would take their daughter home to die. And they proceeded to do just that. The missionary surgeon who would have done the surgery was busy and I was left with the nurses who clearly did not share my distress over the result. To them it was an honorable decision.

So, I said, "Can I talk to you, because I am troubled by this choice?" They said, "Surely." Then, providentially, I asked the right question: "What would these parents have done if this had been a little boy?" After a moment of silence they said simply, "They would have done the amputation." "Why would they treat little boys and little girls differently?" And they gave me the multicultural answer on a plate: "In our culture, it is a woman's job to till the fields, fetch the water, cook the food, and bear the children. A woman with one leg cannot do these tasks and so she will have a life not worth living."

Now if you are a radical multiculturalist, you must accept that understanding of the relative value of little boys and little girls as equally good as a view that dominates Western culture generally. A story like this confronts the radical multiculturalist with the fact that no one truly is a radical multiculturalist. Everybody has to live within an informative story. We all have views influenced by our culture.

What are the implications, then, of deconstructing the radical, unthinking, "all cultures are equal" approach to the modern obsession with making everyone feel good? What follows from accepting that all cultures are clearly not equal? In the case at issue, there is little doubt that the vast majority of people in the Western world would feel that the equal regard for little boys and little girls that has developed over time in our culture is better than the view of young children in that African village. Lest this appear to be blind ethnocentricity, we should realize that those same Africans have many things to teach Westerners about cultural insights in relation to ethics. They, for instance, will not accept that Westerners have homeless working poor. They think, surely a rich nation such as the United States or Canada can help someone build a house. And they are right. Their insight that everyone in the community has a moral responsibility to help the less advantaged is quite right. The fact that there are fifty-thousand-dollar cars running around the same streets where homeless people are forced to live is obscene. Here is an example where the pagan cultural insight is better than so-called Christian societies.

Physicians should be very aware that our own tradition and our own practice of medicine go back, by the grace of God, to Hippocrates, a polytheistic pagan. He understood that the key elements required for the ethical practice of medicine were all moral insights and convictions which required a firm belief in transcendence and divine judgment if they were truly to order the life of the physician. These beliefs in their turn produced a moral ethos in which the sanctity of all life was pivotal. Finally, Hippocrates understood that these truths were so important that the moral integrity of the physician rightly trumps the demands of the patient that they be abrogated for his benefit. He knew these things twenty-five hundred years ago and we ought to be grateful that he did. Today's society has certainly forgotten.

Philosophical Problems with Multiculturalism

Charles Taylor[1] is as succinct as anyone on the philosophical shortcomings of a multiculturalism that thinks liberal democracy can provide a meeting ground for all cultures. Western procedural liberalism (of which more in a moment), Taylor says, "is the political expression of one range of cultures and quite incompatible with other ranges." He uses the Muslim response to Salman Rushdie's *Satanic Verses* and the Quebeçois language laws as examples where procedural liberalism breaks down. Taylor goes on to show that the modern demand for recognition by various marginalized and victim groups is an inevitable consequence of multiculturalism. The irony is that the logic of multiculturalism becomes incoherent and what started as a means to hold us together becomes impetus to fragmentation and divisiveness.

However, if multiculturalism is so wrong, how did it come to be so acceptable? A little history helps, but which history? One view is that most of our problems arose from Christian misbehavior, and that we need to acknowledge this.[2] The misbehavior of Christians that was preeminent in this respect was the Thirty Years' War, which was one political consequence of the Reformation. Catholics and Protestants thought the way to persuade the other side was to beat them over the head and in various other ways cause violence to one another until they saw the "truth." The biography of Johannes Kepler[3] provides vivid and tragic illustrations of how much this religious intolerance disturbed the lives of ordinary folk. The Thirty Years' War was probably one of the most brutal wars fought in Europe and it was brought to an end primarily by the wisdom of humanists who persuaded crusading Christians that, if they could not agree about religion, then religion must be removed from politics. Of course,

1. Charles Taylor, *Multiculturalism* (Princeton: Princeton University Press, 1994).

2. Taylor, on the other hand, starts with Rousseau and argues that it was the need for recognition that started the process. There is a great deal to be learned from a consideration of this viewpoint which space does not allow here.

3. Max Caspar, *Kepler*, ed. C. Doris Hellman, trans. Owen Gingerich and C. Doris Hellman (Mineola, NY: Dover Publications, 1993).

what this meant in practice was that divisive humanly constructed doctrines were to be removed from politics, thereby allowing the development of political peace so that the war could be fought with other nonviolent weapons.

However, there was no way to take acculturated Christian virtues and Christian modes of thought about moral behavior out of politics — a fact to which the distinctively Western political tradition that evolved testifies. This fact was unacknowledged then and is commonly denied now. The solution to the problem of sectarian warfare was the imposition of religious tolerance, which eventually led to the doctrine of the separation of church and state. Initially, the intent was to keep the state out of the church because the church had no formal political power, but it has come to mean that the church has no right to say anything in the public square. That, of course, was not the intention of the American Founders and is nonsense. The church's job is to interfere with the state. Ironically, the church was the first to develop the idea of the secular as separate from the sacred.

Nevertheless, out of this evolutionary process developed a new understanding of the necessity of tolerance for contrary positions — a form of civility for which we should be thankful. We are still learning how to be civil without compromising the practice of vigorous truthful debate. There is still a need to learn that religious dialogue is better than religious war. At the moment it is probably the Muslims who need to learn this more than anyone else, but given that Islam assumes a theocratic state and has a long history of forcible imposition of Islamic government it will be a long time coming. The good news is that the Christian world has, by and large, forsworn violent means of evangelization.

The Hierarchy of Virtues

Tolerance has become the primary, if not the only virtue, of the modern university. What is a "primary virtue"? How do virtues differ between two cultures? The difference between cultures is very much related to the understanding of the hierarchy of goods or virtues in each. What is most important in one culture is not most important in another. Though all the main elements — respect for life, for family, for property — occur in all societies, the relative weight that is attached to each varies.

A good example of this hierarchy can be found in Thessiger's marvelous account of the Horn of Africa, titled *The Desert Sands*.[4] As Thessiger describes,

4. Wilfred Thesiger, *Arabian Sands* (London: Penguin Books, 1981).

the tribesmen considered the property of water so important that if someone took from their well they killed without any compunction or guilt. When the Somalis arrived in North America they brought a remnant set of these tribal desert values — virtues, if you like — which proved to be far from totally compatible with North American society. Tribal societies tend to put loyalty to family, clan, and tribe above truth and above the sanctity of life. This was the problem in Rwanda. Although most Rwandans would call themselves Christians, they did not inhabit a Christian story and their virtues were still pagan, dominated by tribal loyalty. This was made apparent by the fact that people who had gone to church together ended up killing one another when civil war broke out.

Another example comes from my own first trip to Africa. The missionaries with whom we stayed complained that the Christian pharmacists stole from the pharmacy. Elaborate systems of checks were established but all this accomplished was to show that drugs were in fact disappearing. The staff denied having stolen the drugs. Now, from a Western point of view that could not be true. To the Africans, however, it was true. The problem was that although these pharmacists were Christian they still inhabited a pagan story. When their "brother" came to the pharmacy with a prescription but no or insufficient money, the pharmacists saw their primary responsibility as seeing that their "brother" got the drug. If he could not pay, the drug was given to him. The major problem was a different understanding of what the primary responsibilities of a human being in a community are.

One of the reasons that we can speak meaningfully about Christian, Jewish, Muslim, and Hindu societies is that although there is much ethical commonality, the way of expressing the great ideas differs. This is where the idea of a cultural story or worldview is helpful. "Cultural story" seems preferable to "worldview" because it implies that behaviors are not so much acquired by cognition but by growing up with particular people who tell particular stories, which powerfully form the life of the community. Jesus did this. His stories have changed the whole world and are endlessly richer than doctrinal tomes.

Cultural Stories

Stanley Hauerwas[5] describes the idea of the cultural story in a beautiful essay based on the novel *Watership Down* by Richard Adams.[6] In the novel a group

5. Stanley Hauerwas, *A Community of Character: Toward a Constructive Christian Social Ethic* (Notre Dame: Notre Dame University Press, 1981), pp. 9-35.
6. Richard Adams, *Watership Down* (New York: Penguin Books, 1973).

of rabbits seeking a new home experience authoritarian, existentialist, and Marxist warrens before founding their own "human" warren. In more prosaic terms, one can say that all societies have a great narrative that provides meaning for life but also carries with it both potentialities and logical limitations. The Western world was formed by the Bible, the Muslim world by the Koran, the pagan world by the book of Nature and so on. One objection to multiculturalism is that it denies the centrality of these books. It is not inappropriate stereotyping to refer to a Muslim, Christian, Buddhist, or Jewish society. Each has its strengths and weaknesses; no pagan society would discover science because the fundamental cultural construct is magical and the underlying premises of science are thereby made unthinkable. All these societies have virtues like honor in common but the particular expression of honor varies (which is why anthropologists typically say all morals are relative).

Thus bankruptcy in the Western world fifty years ago required liquidation of all assets and the payment of as many creditors as possible. Now we hire lawyers to pay as few creditors as possible. Conversely, in Japan suicide was the honorable action under such circumstances. Honor exists in all three, but its expression varies; anthropologists study only the expression. When these differing stories meet, tensions are inevitable. The multicultural solution requires that all stories be treated as equally true or equally false. The stories are privatized and the public square is allegedly devoid of an informative story. Of course, it is not that we simply have a debased and truncated story substituted for the wealth that was formerly ours. The virtues that formerly required rich allegories, myths, parables, and fairy tales for their propagation are almost all gone. What is left? Tolerance and meaningless alienation.

The modern espousal of tolerance as a central requirement for membership in a pluralistic society presumes to be the solution to an overly strong commitment to our particular understanding of what our own cultural story means. This has some serious problems that need to be unpacked, not the least of which is the alienation of modern deracinated, story-less youth. Tolerance is never without controlling factors, it is not autonomous, it is not independent; it is contingent upon the environment and cultural story in which it is to be expressed. As the story grows thinner, tolerance will easily degenerate into tyranny as pagan norms of loyalty are demanded. Ex-communists and feminists have both described this phenomenon, as in Arthur Koestler's *Darkness at Noon*.[7] One needs to make this point very forcibly, because in our society at the moment there are those who are pushing for accepting as good

7. Arthur Koestler, *Darkness at Noon* (New York: Macmillan, 1941).

what we have been taught by Christianity is evil. This outlook is usually promoted on the grounds that it is intolerant not to accept the behavior that is at issue. Dissident opinions are successfully labeled as "hate speech." In the past, growing up within the biblical narrative made such an idea unthinkable. The removal of the Bible from schools was a key move in the multiculturalist agenda. Without a book like the Bible as an authoritative source, law ultimately becomes the expression of power rather than justice. This trend is evident in the emerging of procedural liberalism.

Procedural Liberalism

By arguing that the law is a means to ensure fair and equal treatment — not a means of translating metaphysics into practical applications — legal theorists have tried to evade the problem of authority. It seems that they have failed, but have not yet acknowledged the failure, and in the interval they proceed as though they are dispensing justice. When multiculturalism removed transcendence from public discussion, it also provided a means for deracinated groups to demand recognition and call for justice defined as equality. Among the most problematic groups are those who define themselves in terms of their sexual desires. If there is no real objective moral truth, then their desires have as much right to be gratified as anyone else's. Thus the demand for acceptance of all sorts of disparate lifestyles is a consequence of procedural liberalism, which presumes that the only function of law is to dispense morally neutral equal treatment while avoiding debate about the ultimate nature of good and evil — and even justice. The assumption is that the law can function without theological or religious underpinnings. It cannot; but there is not space to discuss the issue here beyond referring the reader to A. A. Leff's[8] classic paper. As Leff points out, without a lawgiver who is truly above us, we are inevitably tempted to ask, "Why should I treat everyone equally?" and "Who says I ought to?"

Looking at justice as merely procedural fairness now dominates our society. Even many evangelical Christians, in their deep commitment to individual liberty, see no problem with this approach despite its deeply atheistic assumptions. It is very difficult to fight an enemy who has outposts in our heads! Many of the ethical problems in science, medicine, and elsewhere arose because Christians in public policy positions thought in post-Enlightenment procedural and utilitarian terms rather than starting from

---

8. A. A. Leff, "Unspeakable Ethics, Unnatural Law," *Duke Law Journal* (1979): 1233.

substantive principles. This was wonderfully illustrated by a comment of Charles Malik, one of the framers of the UN Human Rights declaration. When asked how he persuaded so many philosophically-opposed groups to sign a common declaration he said, "As long as you don't ask why we agree, it is all right."

How do we deal with this? Logically, unprincipled government will eventually result in corruption. We have to learn to confront these issues, to help others face the fact that they do actually believe in real moral truth. A good way to open up the debate is to ask questions rather than make statements — to propose thought experiments in which one's own particular position is not actually revealed but is, in fact, expressed. One example is the following.

Most people do not know what NAMBLA means. It is the acronym for the North American Man Boy Love Association — an association that exists for the purpose of legalizing sodomy between adult men and young boys. Its slogan used to be "Eight is too late." Consider the following thought experiment. Imagine that you are parents of an eight-year-old boy and you have a militant member of this association as a houseguest for three or four weeks. The man turns out to be a charming young man who is creative, reads to and plays with the kids, and even cooks. He is one of the best houseguests you have had, though you have qualms about his statistically unusual view of normal sexuality. Are you going to allow this charming and sophisticated thirty-year-old to persuade your eight-year-old that he is missing out on some of the rights of all eight-year-olds? Of course, your houseguest would never wish to fulfill his desires without permission. Would you give permission? I have never found anyone who answered "yes."

So I must say to them, "Welcome to the ranks of the intolerant. There is something which you will not tolerate." And now we get to the really important issue: is that intolerance justified? Yes, it is justified on the grounds of love. Parents have no reason to be in any way intimidated, for their understanding of what would be loving for that boy is rooted in eight years of sacrificial service, and their houseguest's is rooted in three weeks of play. In fact, we can go a step further and say that whenever there is an attack on a primary good like love or truth or honor, it is actually our duty to be intolerant of that attack. To be silent is to be foolish. It is also, according to Dorothy Sayers, a failure to appreciate what the radical elevation of tolerance to first place would do. Her description of tolerance is quite challenging:

> The church names the sixth deadly sin Acedia or sloth. In the world it calls itself tolerance, but in hell it is called despair. It is the accomplice of the

other sins and their worst punishment. It is the sin which believes in nothing, interferes with nothing, enjoys nothing, finds purpose in nothing, loves
nothing, hates nothing, . . . and only remains alive because there is nothing
it would die for.[9]

This description rhetorically overstates the case, but it is a necessary correc-
tive for today's failure to properly understand the nature of appropriate intolerance. Perhaps we should rename the Ten Commandments the Ten Intoler-
ances. Even those who pride themselves on their nonjudgmental tolerance are
likely to support Amnesty International and to expect truthful speech from
their colleagues.

The Necessity of Judgment

If tolerance is inadequate as a basis for civil society and its history explains
how it came to be so dominant, what must we do now to see it confined to its
proper usage? We must say that tolerance is very important wherever there
are uncertainties in our knowledge. The Protestant emphasis on our human
fallenness is a good starting point. We do need to be careful not to take our
own cultural background as somehow truer than another without actually
going through the effort of working out whether it is, as in the case of the septic knee. Moral neutrality will not work.[10] To assert moral neutrality is in fact
to make a judgment; it is to judge that we should not judge. We cannot avoid
judgment. It is a part of life. As Wittgenstein said, "Ethics is a condition of
man." We have to judge every day. The question is on what basis do we judge,
what model do we use? That there are different stories out there is clear. Our
job is to understand those stories and then apply them to the situation in
which we find ourselves to show which one works best.

We are currently — in the political arena — at a point where a Christian
understanding of good and evil is beginning to assert itself in a more articulate way. We have no reason to be ashamed of this position. Not only can we
say that all injunctions about how we ought to behave are ultimately based on
faith but, as Robert George has eloquently outlined, we can say that on purely
rational grounds, Christian understandings of the sanctity of life, marriage,

9. Dorothy Sayers, "The Other Six Deadly Sins," in her *Creed or Chaos* (Manchester,
N.H.: Sophia Institute Press, 1995).

10. A more detailed discussion on moral neutrality can be found in John Patrick's
essay, *The Myth of Moral Neutrality* (Warkworth: Christian Medical and Dental Society,
1997).

and sexuality produce better outcomes.[11] Evangelical Christians in particular have new reason for encouragement. In the past century, they have retreated into pietism, pentecostalism, and dispensationalism, which are in various ways incomprehensible to the university elites. The result has been the exclusion of evangelicals from meaningful discussion with the intellectual and political elites who rule today's world. Fortunately, this phase will soon be past, in part because evangelicals are again contributing to the intellectual and cultural war which is now under way, and also because the multiculturalism adopted by the universities and government attempts to treat all groups with respect.

No Christian who believes that Jesus is the Son of God and that his death is the means of our reconciliation with God can subscribe to the notion that all cultures are equal. Until now, Christian convictions have been dismissed as merely personal faith. As this discussion is subjected to closer scrutiny, however, it is increasingly clear that to rule as though God, if He exists, is irrelevant to the process of government is not a tenable position. It is to impose a tacitly atheistic belief system. The fundamental problem is that the real moral truths that are under attack are so deeply buried that we do not meditate on them; we simply live by them. All that we are aware of is a vague disquiet in our souls when they are attacked. That the attack is subtle and claims to be motivated by concern for the oppressed merely makes it that much harder to repulse.

The progressive loss of a moral consensus in the Western world over the last two centuries is accelerating. At some point the breakdown will be such that we can no longer agree. MacIntyre's *After Virtue* is very important in this regard.[12] After a magisterial review of Western cultural history, MacIntyre concludes by arguing that we have now entered upon a second dark age. We will, at some point, have to cease the attempt to shore up increasingly corrupt government and put our efforts into the formation of communities within which we can maintain the virtues and civilities which are the heritage of Christendom. As a community Christians are already doing this in education and the results are increasingly impressive. Medicine and insurance may well be next.

---

11. Robert George, "The Clash of Orthodoxies: An Exchange," *First Things* 104 (June/July 2000): 45-52.

12. Alasdair MacIntyre, *After Virtue* (Notre Dame: Notre Dame University Press, 1984).

Reliance on Technology:
Stem Cell Research and Beyond

DARYL SAS

Few people would doubt that technology has had an enormous impact on health care. At the same time, however, most would add that technology has also introduced serious ethical problems. For example, *in vitro* fertilization (IVF) is now commonplace and accepted by many people, but once was a technological breakthrough. Remember the public reaction to the first human "test tube" baby? Accompanying technology has given us the ability to freeze and preserve embryos created *in vitro,* raising the ethical dilemma: what should we do with those who are not implanted? A second example is cloning. In the last three years, sheep, mice, cows, and pigs, among other animals, have been cloned. How soon will it be before a human is cloned?

One further example is embryonic stem cell transplantation. This technology promises to treat Parkinson's, diabetes, arthritis, heart attacks, autoimmunity, osteoporosis, cancer, Alzheimer's disease, severe burns, spinal cord injuries, birth defects, muscular dystrophies, multiple sclerosis, strokes, and liver disease, among other conditions — potentially helping a total of 128.4 million people.[1] How could anyone, including the federal government, deny this benefit to so many people? Yet the embryos destroyed in the process are human beings who should not be exploited for the supposed "good" of others. Without a doubt, technology will remain just as important in the next millennium, continuing to run ahead of our ability to consider thoughtfully

---

1. Daniel Perry, "Patients' Voices: The Powerful Sound in the Stem Cell Debate," *Science* 287 (2000): 1423.

its applications, unless we act promptly to set boundaries for technology based on the only truly objective source of ethical principles available — God's word.

The Appeal of Technology

Before we can identify relevant biblical principles, we need to understand the appeal of technology, especially to our secular culture. Technology is not something negative, in and of itself. It simply must be relegated to its proper place: subject to God. Technology's charms are considerable for those without faith in God. However, while technology can be attractive, its long-term effects can be hidden by its initial benefits. Its charms need to be scrutinized before we as a society are lulled into missing the eternal, ethical implications.

First, technology promises to free us from the pain and suffering inherent in a fallen nature: new antibiotics for infections, better Doppler radar to warn us of tornadoes, and more earthquake-proof buildings. It offers freedom from the limits imposed by society, morality, and our human bodies. With online trading, anyone can become rich regardless of societal background. Vaccines for herpes, human papilloma virus, and HIV will protect us from such serious conditions, whether or not they are God's judgments. We now have "labor-saving" technologies, such as road construction equipment to enable us to avoid some of the work and sweat inherent in the creation cursed by God (Genesis 3:19). We also have herbicides to combat the thorns and thistles resulting from God's pronouncement to Adam and Eve (Genesis 3:18). The penultimate judgment, of course, is physical death (Genesis 3:19); but cryopreservation and cloning, according to their proponents, promise to allow us to elude death as well.

Second, technology promises to allow us to be fully human, at least by humanism's definition of humanness: with total freedom and independence. According to this definition, the freer we are, the more human we become. Such freedom enables us to realize all of our human potentials and have all kinds of human experiences. Technology purports to make us fully human by providing freedom from external constraints. For example, if a lack of access to information is stifling children's development, they have only to log onto the Internet. Another tenet of humanism is that the less dependent we are on others, the more human we become. In this respect, physician-assisted suicide makes us freer to "go" where, when, and how we want to. "Death with dignity" is, in this view, controlled death. Conversely, of course, the more dependent and out of control we become, e.g., hooked to a respirator and confined to a hospital bed, the less truly human we are.

Third, technology promises happiness, which, without God, basically boils down to comfort and physical pleasure. Technology claims to ensure economic prosperity, which is closely connected in many people's minds with happiness. Technology also offers choices, which in turn add to the freedom that so many desire as a source of happiness. Of course, the happiness that technology brings is not a lasting, spiritual happiness or biblical joy, but momentary comfort, like fast food, air conditioning, or headache relief.

Often, technology only offers diversions — usually brief thrills to distract us from mind and body. Think, for example, of video games or virtual reality. Technology offers an escape from the serious, meaningful contemplation of being human: mortal in body but eternal in spirit. Preoccupation with such technology is not so much ignorant bliss as it is self-deception.

Ultimately, technology promises salvation. Since the old remedies for our problems such as old antibiotics and old insecticides have lost their efficacy, new technology is our only recourse. Since God, if He is even there, seems too slow to act, science will take care of our problems; technology will save us from problems with disease, aging, and fertility.

Technology, then, can provide a way to replace God. Technology is given attributes of God. Technology is seen as uniquely capable, almost holy; powerful, almost omnipotent; leader, almost lord. Has technology, in the eyes of some, already replaced God? If we want to know who someone's god is, we should look at whom they worship and praise, who demands the most of their time and energy, whom they trust for deliverance and what they call paradise. We often hear people in secular settings praising technology, and looking to it for truth, as well as assuming it will save them from their problems and bring them into utopia. Are these the "offerings" demanded by this new god?

Two sets of promises compete for our trust: those of God and those of technology.

The Technological Bluff

The French theologian and philosopher, Jacques Ellul, in his book *The Technological Bluff*, argues convincingly that technology's promises are exaggerated.[2] It is interesting that a recent editorial in *Science* magazine agrees by beginning with the statement, "The Human Genome Project has engendered

2. Jacques Ellul, *The Technological Bluff* (Grand Rapids, MI: Eerdmans, 1990).

genohype. . . ."[3] Such exaggeration is also involved in the panacea claimed for human embryonic stem cells, as will be explained later in the chapter. Ellul accuses the proponents of technology of concealing its risks. This seems to be a fair accusation. Do the commercials for online trading clearly reveal the risks involved? Might the same accusation be leveled at promotions for gene therapy?

A second problem with technology is that rather than getting at the causes of our problems, it all too often only masks their symptoms. This would certainly seem to be true for diseases which are caused by certain behaviors, such as alcoholism, smoking, or sexual promiscuity, which by some estimates produce nearly forty percent of illness. But technology also masks — and keeps us from coming to grips with — our ultimate problem: that "the wages of sin is death" (Romans 6:23).

Ellul points out that the critics of technology are not welcome. Pessimistic opinions are rejected because they are not what people want to hear. People want easy hope or a quick fix. Most people value the immediate benefits of comfort, control, independence, and autonomy over the distant risks of deformed babies, infertility, or eternal judgment for tampering with or destroying human embryos. People much prefer the illusion of pulling themselves up by their own bootstraps over the reality of dependence on God for salvation. C. S. Lewis, in his essay "The Funeral of a Great Myth," suggests that this is the allure of evolutionism: the ability to save one's own species, if not one's self.[4] Salvation and ultimately perfection are possible without God, some say, through evolution. Others would apply this same concept to technology. Of course, with the complete sequence of the human genome in hand — another technological *tour de force* — even evolution will supposedly be under our control. What is the place of God in all this?

A Biblical Perspective on Technology

A fully-developed Christian perspective on technology is beyond the scope of this chapter. Nevertheless, there are some biblical insights that can be very helpful as we begin the task of developing such a perspective, for the word of God is indeed "a lamp to my feet and a light for my path" (Psalms 119:105).

3. Neil Holtzman, "Are Genetic Tests Adequately Regulated?" *Science* 286 (1999): 409.

4. C. S. Lewis, "The Funeral of a Great Myth," in *Christian Reflections*, ed. Walter Hooper (Grand Rapids, MI: Eerdmans, 1967).

What does the Bible suggest about technology? First, technology seems to be a valid response to the mandate to "fill the earth and subdue it" (Genesis 1:28). "Subdue" has generally been interpreted to mean: investigate and understand, control and direct, and care for and develop. Technology certainly seems to facilitate such endeavors.

However, it is telling that, in biblical times, technology was largely the work of unbelievers. For example, the Egyptians built chariots and pyramids. The Hittites were the first to make iron. The Babylonians constructed ziggurats and hanging gardens. The Greeks engineered new fighting ships and advanced architecture. The Romans designed aqueducts, heated baths, and advanced weapons. Why such emphasis on technology by the pagan Egyptians, Canaanites, and Gentiles? For the same reasons that technology is emphasized in our time: it offered them salvation from temporal enemies, control over their futures, comfort amidst pain and suffering, and a name — a lasting reputation, the desire for which hearkens back all the way to the Tower of Babel: "Then they said, 'Come, let us build ourselves a city, with a tower that reaches to the heavens, so that we may make a name for ourselves and not be scattered over the face of the whole earth'" (Genesis 11:4).

What was God's response to the technological advances of the pagans? God delivered salvation to the people of Israel from Pharaoh and his chariots by destroying those chariots. God displayed His sovereignty over Nebuchadnezzar's independence and supposed autonomy by removing his authority and making him like cattle until he would acknowledge God's greatness and sovereign rule over people. In doing all this, He reminded the nations that He is holy, omnipotent, Lord and Deliverer. God's people also had to learn this lesson again and again. They employed technology to a degree — often as a means of carrying out God's will. But how easy it was — and is — for that technology to give people the illusion of power and control, and the idea that God is unnecessary. King Uzziah is a case in point:

> Uzziah provided shields, spears, helmets, coats of armor, bows and slingstones for the entire army. In Jerusalem he made machines designed for skillful men for use on the towers and on the corner defenses to shoot arrows and hurl large stones. His fame spread far and wide, for he was greatly helped until he became powerful. But after Uzziah became powerful his pride led to his downfall. (2 Chronicles 26:14-16)

The problem is not technology per se. Technology can provide the means through which God brings great blessing. But we must not lose sight of God's presence at work through it, and consequently forget the importance of holding it accountable to God's standards. "For from him and through him and to him are all things" (Romans 11:36). Do "all things" include technology? Yes. Our ability to create technology is ultimately from Him. God is the source of the raw materials which technology uses, the source of the laws governing His creation, and the source of human creativity and rationality. He can work through technology, but He can also work independently of it.

Technology should ultimately be evaluated in the context of our service to God. God is, or at least ought to be, the ultimate recipient of the use of technology. It is for His glory. It is for the building of His kingdom. It is for the hallowing of His Name. It is for the benefit of His creatures — "I tell you the truth, whatever you did for one of the least of these brothers of mine [directly or indirectly, by using technology or not], you did for me" (Matthew 25:40). Technology directed to other ends is essentially idolatry. God demands that we love Him with all our heart, all our mind, all our soul, all our strength . . . and all our technology.

But these wonderful ends do not justify the means used to obtain them. Technology must also be done "through Him." That means, in part, that God sets the boundaries for the ethical use of technology. In His word, God has given three standards by which to judge the ethics of an action: the action must be obedient to biblical law, it must be motivated by biblical love, and the consequences must be measured by biblical justice. Notice how this biblical view integrates the three general theories of ethics — deontological, virtuistic, and consequential — and defines each one biblically. Let me underscore: an ethical use of technology would be obedient to biblical law, motivated by biblical love, and measured by biblical justice. For a somewhat different formulation of these biblical standards, see John F. Kilner's *Life on the Line*.[5]

As one example of the practical application of these guidelines, let us examine a controversial new medical technology — the transplantation or other use of human embryonic stem cells (ESCs). This technique harvests the primordial stem cells present in embryos that may be no more than 32 to 128 cells in size. These cells are particularly desired because they can supposedly differentiate into any type of human tissue. Unfortunately, the embryo is destroyed in the process. Controversy has arisen over whether the process is ethically acceptable at all, or if it is, which embryos should be used. The President's National Bioethics Advisory Commission has suggested that embryos that have been conceived and cryopreserved in the process of in vitro fertilization treatments for infertility could be used if there was no longer any

---

5. John F. Kilner, *Life on the Line: Ethics, Aging, Ending Patients' Lives, and Allocating Vital Resources* (Grand Rapids, MI: Eerdmans, 1992).

expectation that they would ever be implanted and allowed to mature.[6] The argument is clearly utilitarian: they are going to be destroyed anyway, so why not produce the greatest good for the greatest number by sacrificing them to treat various diseases? What is a Christian response to this?

First we must ask if such use of human ESCs is obedient to biblical law. One key biblical standard involves the reality of who the embryo is. If the embryo is a human being,[7] ESC research is the equivalent of murder — a clear violation of the sixth commandment "You shall not kill" . . . even embryonic persons. To claim that the life is going to be lost anyway is not a sufficient defense. It is like saying "these Jews are going to be gassed anyway, so let's do horrible experiments on them to gain information that will benefit our soldiers" — precisely the reasoning many Nazi physicians used to justify many of their atrocities. Rather than exploiting an unfortunate problem, the biblical mandate to "choose life" would dictate that means should be found to allow those unwanted embryos another "home" where they will be given a chance at development and life.

Is the use of human ESCs motivated by biblical love? We do have a biblical mandate to care for the afflicted. But doing so must be guided by biblical love, which is defined as selflessness. Caring for those afflicted by disease should never be at the expense of others' lives. We would never tolerate murdering someone for their liver, so neither should we allow the killing of an embryonic person for his/her stem cells. While we may selflessly sacrifice our own lives for the benefit of others, we do not have the right to sacrifice others' lives for our own selfish gain.

Finally, is the use of human ESCs just? Biblical justice demands that the embryo's need for life outweighs the patient's need for comfort or freedom, regardless if it is freedom from the pain of arthritis, or the limitations due to heart disease, the crippling effects of osteoporosis, etc. All such consequences are secondary to the embryo's need for life. It is not just to deprive one person of life to purchase something of lesser value for another.

So we are driven to the conclusion that on all three biblical grounds, this is an unethical use of technology. Even if it violated only one of the three guidelines, we would have to label it unethical, or at least less ethical than other alternatives that follow all three guidelines. The same approach ought to apply to whatever medical technologies arise in the new millennium.

6. The President's National Bioethics Advisory Commission, *Ethical Issues in Stem Cell Research*, Vol. 1: *Report and Recommendations of the National Bioethics Advisory Commission* (Rockville, MD: 1999), p. 70.
7. Robert W. Evans, "The Moral Status of Embryos," in *The Reproduction Revolution*, ed. John F. Kilner, et al. (Grand Rapids, MI: 2000), pp. 60-76.

Loving regard for the needs of the sick, the orphaned, and the widowed compels us to do all we can within ethical limits — and there is a great deal that can be done. Scientists have extensive evidence in the last two years that adult stem cells can be obtained and manipulated to form other tissue types.[8] Though further work is required, it appears that embryonic stem cells will not turn out to be any more useful than adult stem cells, and may, in fact, be even more recalcitrant.[9] Not only will loving concern for those afflicted with disease not require the destruction of "orphaned" embryos, but such love actually commends the pursuit of adult stem cell research over embryonic stem cell research. Adult stem cells taken from the person who is ill might be used to produce healing tissues that are an exact genetic match to the tissues of that person, thereby minimizing the risk of painful and costly rejection of that tissue by the person's immune system.

No one need fear technology, but we need to approach it wisely, remembering that "the fear of the LORD is the beginning of wisdom, but fools [the morally deficient] despise wisdom and discipline" (Proverbs 1:7). We need to evaluate new developments critically, not accept them unthinkingly or be seduced by their claims. And as we consider new technologies on the horizon, we need to evaluate them prospectively and develop biblically-based guidelines for their implementation. In this way we will be good stewards of the material gifts and creativity God has bestowed upon us.

8. T. R. Brazelton, "From Marrow to Brain: Expression of Neuronal Phenotypes in Adult Mice," *Science* 290 (2000): 1775-1779.
9. Gretchen Vogel, "Stem Cells: New Excitement, Persistent Questions," *Science* 290 (2000): 1672-1674.

The Need for Bioethical Vision

FRANCIS CARDINAL GEORGE

The announcement that the federal Human Genome Project has succeeded
— ahead of schedule — in mapping the human genetic code convinces me
that we have arrived at the threshold of the Enlightenment project's final
frontier. At the core of this modern project, a project not totally unrelated in
its genesis to Christian values, is the notion of the intrinsic value of each human person, and the subsequent desire to *liberate* all persons, to free them
from the unfulfilling strictures of other persons and even from those limitations that physical nature imposes through the human body. According to the
secular version of modernity, the *means* to this freedom and fulfillment for all
are the free economy, the liberal democratic state and, above all, the scientific
method of sensory observation and inductive logic that reveals the material
and efficient causes operative in both physical and human affairs.

This combination of science and liberalism has recently generated a new
global phenomenon and a corresponding new paradigm for describing the
way in which the human family can and should relate: "globalization." The
"Cold War" world that was formerly *divided* into two competing perspectives
on the human person and the social order has gladly passed, giving way to a
new situation in which the human family is not only united but is ever more
interconnected. While interconnectedness is hardly new to human history, we
can now more easily, rapidly, and cheaply than ever move — and thus share —
ourselves, our consumer goods, our material and human capital that together
produce goods, and the values that comprise our respective cultures.

The two developments responsible for this interconnectedness are two
manifestations of the scientific and social aspects of modernity: technology
and liberalism. Communication and transportation technologies have made

it possible to move information, ideas, people, and things more easily. More than at any time in human history, the *physical* constraints of time and space are amenable to our desire and will to transcend them. These technologies, however, would not carry our ideas, people and things without an agreement that such exchanges are beneficial and/or right. This agreement — called "the Washington consensus" in some circles — is the result of the arguments of economic and political liberals who together stipulate that economic freedom is conducive to human flourishing. In the case of globalization, it is argued that technology and economic freedom can together promote more plentiful and cheaper products that reflect consumer preferences and needs, international development and solidarity, liberal democratic regimes, international peace, and positive cultural characteristics. Some of these arguments are convincing, both logically and empirically — but not all.

It is also becoming increasingly apparent that because of human moral limitations, or sinfulness, our increasing interconnectedness also holds the potential for offenses against human dignity, and that too many of the choices afforded by globalization are lacking in respect for others and their natural and cultural environments. These choices need to be better directed — culturally and politically as well as morally — in order to both promote authentic human development and avoid political backlash. The goal, as Pope John Paul II expresses it, is globalization *with solidarity.* Interconnectedness is not automatically relationship; it is relationship to God through Christ, and to one another in Christ's body the Church, which is salvific or liberating from the viewpoint of Christian faith.

However, the ability to transcend the constraints that time and space impose upon the body (which is one of the conditions for the possibility of global interconnectedness and therefore a basis for developing relationship) is *not* the final frontier of the modern project to maximize individual freedom. *The final frontier is the body itself.* In Pope John Paul II's words, "The human genome in a way is the last continent to be explored."[1] Especially over the last five hundred years, advancements in techniques and theories have thankfully produced a revolutionary increase in our knowledge of the way the body works, and of how to prevent and treat its maladies. Francis Bacon's recommendation at the beginning of the Enlightenment project that we "put nature to the test" in order to "relieve man of his estate" has been heeded with stunning results. One result of this enterprise to understand and control nature has been the discovery of the genetic basis for personal characteristics — physical, mental, and emotional. We are still far from completely knowing

1. Address to Pontifical Academy for Life, February 24, 1998.

which nucleotide sequences correspond to these traits. We are even further from completely knowing how to manipulate chromosomal sites in order to promote human fulfillment. But we already know the genetic basis of many characteristics, and are beginning to develop intervention techniques that allow us to alter the genetic code of both somatic and germline cells.

As with the globalization process, this knowledge has the potential to contribute greatly to authentic human development. In his 1983 discussion on the *Dangers of Genetic Manipulation,* John Paul II lauded the advent of therapeutic genetic intervention on somatic cells — as long as these procedures are governed by ethical considerations similar to those of tissue or organ transplantation. These techniques must first be developed using animals; then their risks — including public health risks — and benefits must be weighed. In these cases, what is ambiguously called genetic *engineering* is actually genetic *therapy* that alleviates unnecessary suffering and restores normal bodily functioning. In the Pope's words: "A strictly therapeutic intervention whose explicit objective is the healing of various maladies such as those stemming from deficiencies in chromosomes will, in principle, be considered desirable, provided it is directed to the true promotion of the personal well-being of man and does not infringe on his integrity or worsen his life. Such an intervention, indeed, would fall within the logic of the Christian moral tradition."[2]

On the other hand, as we see with the globalization process, and as we have seen with all scientific and technological advances, our increasing knowledge and ability to transcend physical constraints also hold the potential for offenses against human dignity and flourishing. Of obvious concern are abuses against unborn human persons — scrupulously called "embryos" or "fetuses" in an attempt to mask their humanity — committed for a variety of purposes, including genetic research and the harvesting of tissue-producing stem cells. The ability to create a clone for one's personal use will constitute the height of self-centered narcissism and the instrumentalization of another. Also of concern are artificial methods of conception, especially in vitro fertilization and artificial insemination using both spousal and non-spousal gametes, that, irrespective of whether the intended end is therapeutic or nontherapeutic, run the risk of destroying human life and promoting a utilitarian attitude toward one's spouse and children. Additionally, while improved prenatal genetic testing and therapies have the potential to enhance life, such tests will increasingly be occasions for both voluntary and coerced abortions based on a misguided biological reductionism and determinism and on an ever-narrowing conception

2. John Paul II, *Dangers of Genetic Manipulation,* October 29, 1983.

of what is "desirable." This passing century has seen what happens when the individual right to life is conditioned by whether or not one is wanted — by a party or a state or one's parents.

For those of us lucky enough to be brought to term, these same factors — the capacity to diagnose predispositions to physical and even behavioral traits, biological reductionism and determinism, and an eclipsed sense of the value and rights of human beings in any condition — will threaten us with "genetic profiling" and the subsequent denial of rights to health care and jobs. In such a hostile cultural climate, the specters of assisted-suicide and euthanasia haunt even those showing the first signs of genetic diseases that are otherwise temporary, treatable, or manageable. Finally, in such a climate, it is not inconceivable that even those who are sick or disabled by pathogens or accident will be seen as genetically responsible for their conditions, and thus be considered as unnecessary burdens, to be marginalized and eliminated. All of the present schemes for extending universal health care by rationing coverage of some diseases threaten to leave large numbers of disabled persons more vulnerable than ever and are resented and feared by them and their care givers.

Gradually, however, the need to abort and euthanize such "genetically challenged" "undesirables" may diminish somewhat because of our increasing ability to create human beings not only free from disease but also in possession of high degrees of characteristics that are indisputably good — beauty, intelligence, affability, longevity. Initially, the means to these ends will be gamete profiling and selection followed by artificial conception. Eventually, however, even these artificial means of conception might be bypassed through germ cell interventions aimed at tailor-making one's sperm and eggs.

Finally, the issue of *non-therapeutic* genetic manipulation now looms before us. What we are talking about here, of course, is *positive eugenics* accomplished not through coercion, but through *free choices* made possible through a free market — although one can imagine also the possibility of coerced *therapeutic* manipulations aimed at limiting the social costs of disease. Aside from unintended consequences such as impairment or death or the intentional production of a useful "lower class" of human beings, what would be wrong with freely choosing to maximize a host of traits that are indisputably *good?* Why not develop now what Christians waiting for the resurrection in faith would recognize as a semblance of the risen body? The risen Christ is not constrained by the laws of physical nature; his is a body that is incorruptible. He is completely free.

The Physical and the Spiritual

Like us, Christ was truly a human being because he was born of a human mother, Mary. From Mary's womb he emerged a person, fully human, like us in all things but sin, and with a clearly physical body. From the Garden tomb he also emerged a man, but no longer entirely like us in his body. For, having passed through a fully human life and the ultimate barrier, death itself, he emerged transformed, possessing what St. Paul calls a "spiritual body." Truly his, but transformed in ways that Scripture does not detail.

In the first letter to the Corinthians, St. Paul says that God provides all living things — plants, animals, humans, and even the heavenly bodies — with the kind of body appropriate to them. Even death does not prevent this: even the apparently dead seed in the ground rises in a new, leafy, and fruitful body. Paul applies this insight to the human body resurrected from the dead: "What is sown in the earth is subject to decay; what rises is incorruptible. What is sown is ignoble; what rises is glorious. What is sown is weak; what rises is strong. A natural body is put down and a spiritual body rises."[3]

There is a technical distinction made clear by Paul's vocabulary describing natural and resurrected bodies. He describes the body both as *sarx*, which is the element of limitation and corruption in our bodies, the result of sin, and as *soma*, a more neutral term which indicates a body capable of being transformed into spirit while remaining a truly material body. The body's principle of life in our time and space is *psyche*, or soul. Its principle of life in its risen state is *spirit*. The risen Christ has become a life-giving spirit, Paul says, in whose likeness believers will be raised. Limitation and corruption cannot inherit the Kingdom of God and, at the Last Judgment, *sarx* is destroyed and *soma* becomes incorruptible and immortal. The nature of spiritual bodies, other than the certitude that they are human bodies transformed, is not revealed. According to the perspective of Scripture and the experience of wise visionaries, light is the metaphor which comes closest to grasping this transformation. Modern studies of light as energy give those who want to use science to pursue this question of faith an important opening for responsible speculation.

If, according to Christian faith, Jesus is the firstborn of the dead, then it is the history of his body that tells us the meaning and the nature and the destiny of ours. The body is integral to salvation history because Jesus our Savior has risen from the dead. An origin in time does not therefore demand that salvation means an escape into eternity. The bridge between *psyche* and spirit

3. I Corinthians 15:42-44.

is that transformation that evangelists and doctrinal theologians call resur-rection from the dead. Just as natural life is a gift, so is risen life pure gift. But between the one and the other comes the crucifixion of the body — Jesus' act of total self-sacrifice made possible, in part, by his material human body. In the light of faith, the gift of life must be surrendered, willingly sacrificed, so the gift of eternal life can be received.

It is in the face of the paradox involving the incontestable superiority of the resurrected and unconstrained body that the value of a distinctively *Christian* bioethical vision becomes apparent. In sketching the outlines of such a vision, however, a few words about the basic relationship between faith and reason are in order, both being necessary for the development of ethical principles. What is the relationship between Christian revelation and both philosophy and science? As many of you well know, Pope John Paul II, him-self a philosopher, has offered the Church a long reflection on the relation-ship between faith and philosophy in *Fides et Ratio.* Throughout this docu-ment, the Pope insists upon the need for purely philosophical methods and conclusions — even contemporary ones such as phenomenological method — capable of both interpreting the data of revelation and making it more in-telligible and thus more credible. Faith is pure gift. But faith according to St. Paul has a double dimension. Faith means trust; and faith means truth. To see the truths of faith, to see the world with the eyes of faith is a gift from the God we believe in. But this same God also gives us the gift of reason, and reason can explore the vision of faith on its own terms in theology and, in philoso-phy and apologetics, reason can prepare a thinker to receive the gift of faith, without ever explaining the mysteries of faith themselves. Theology and faith need philosophy and human reason and science. The Pope is clearly opposed to a fideism understood as an irrational faith.

On the other hand, he also insists that reason and philosophy and sci-ence as well need theology. He is opposed to a rationalism that limits reality to what human reason can know. Through the people of Israel, through Christ and his Church, God has revealed truths that both purify reason and compensate for its limitations.[4] Grace, in other words, "does not destroy na-ture but perfects it."[5]

The relationship between faith and modern science, a relationship with a rocky past, continues to be misunderstood. John Paul II's first major act as Pope to deepen the dialogue between the Church and science was his 1979 ad-dress commemorating the centenary of Albert Einstein's birth in which he ac-

4. John Paul II, *Fides et Ratio,* 75.
5. Ibid., 76.

knowledged "the greatness of Galileo," lamented his treatment "at the hands of churchmen and Church institutions," and called for a more "fruitful concord between faith and science." In 1987, marking the three hundredth anniversary of Newton's *Principia Mathematica*, he stressed the complementarity of theology and science, and the need to integrate the insights of both in order to understand better the human person. In his words: "No one can read the history of the past century and not realize that crisis is upon both [theology and science]. The uses of science have on more than one occasion proven massively destructive, and the reflections on religion have too often been sterile. We need each other to be what we must be, what we are called to be."

These themes were revisited in 1992 on the 350th anniversary of Galileo's death. The Pope received a report from the Pontifical Council for Culture indicating that the Church had erred in associating the faith with "age-old cosmology," in believing that "the adoption of the Copernican revolution . . . was such as to undermine the Catholic tradition." The Church needed to recommit itself to the ancient Christian tradition that all knowledge which sheds light upon the human person that the Church exists to serve is welcome. Scientific truths — such as the replacement of an Aristotelian physics of rest with a Newtonian model, heliocentrism, and microevolutionary development — cannot contradict the truths of the faith, and may actually assist us in interpreting the meaning of revelation.

Science, on the other hand, cannot operate completely independent of the kinds of normative considerations that come from a faith rather than from science itself. Underlying the scientific method are concepts that transcend the empirical, such as the existence of a law-governed material world intelligible to human reason. In this context, the modern scientific project is properly seen as grounded in a number of beliefs. Also, and perhaps more importantly, science, as a purely descriptive enterprise, cannot provide an account of the human good, and therefore cannot provide answers about which effects we should cause and which means to them are licit. Such *normative* considerations are the domain of both philosophy — specifically, a metaphysics of the person and ethics — and theology, which, as indicated earlier, should guide philosophical reasoning.

Unfortunately, however, our *culture*, in granting epistemological hegemony to the scientific method, has not only contradicted science's theologically-supported foundations, but has also rendered meaningless any universally valid normative claims about such issues as the value of persons and what constitutes authentic human fulfillment. The story goes something like this: since known states of affairs — "facts" versus "values" — are those that we sense or measure, and because we sense in human affairs a plurality of conceptions of human value and flourishing, there are, logically, no universally valid normative concepts. According to such a secular humanism, the person is reducible to the material body, the condition of which is largely determined by his or her genes. In the absence of a philosophically or theologically derived anthropology that would guide human action, the goals of relationships, economics, politics, and science tend to become the maximization of the body's pleasure or the achievement of individual purpose. In practice, of course, this means maximizing *one's own* pleasure or purpose even at the expense of others, since "the other" has no indisputable, intrinsic dignity, and since a life of self-giving is only one "lifestyle" among other options. Hobbes' assertion that there is no "highest good" does indeed translate into "a war of all against all."[6] In this war, the weak — the unborn, the poor, the disabled, the genetically "inferior" — are the inevitable losers.

Ultimately, if guided only by empiricism and relativism, the modern project risks undermining its own stated goals. Even if we accept the Kantian notion of a "moral imperative" that imparts a value to the other (specifically, to "rational beings," who become "persons" as opposed to "things"),[7] without agreeing upon a conception of human flourishing, we are left without a criterion for determining what constitutes respect for others. Differently stated: absent a shared notion of human flourishing, *which rights* — the conditions necessary to flourishing that are to be protected against denial by others — are due?

A Christian *bioethical* vision should thus be grounded in a Christian *anthropological* one, in the truths about the human person that revelation — especially the risen Christ — discloses and reinforces. First among these is the dignity or value of each human person at every stage and condition. Even without the aid of revelation, the uniqueness of each human being, the implications of natural law precepts and the intentional affective response to persons commonly called "love" would all recommend this truth to us. But additionally, the revelation of our likeness to and relationship with God, especially as revealed through the Incarnation and Christ's self-sacrifice, incontestably attest to this truth. The God in whose image each of us is created "knows" and "consecrates" us in the womb, and sent his only begotten Son, his Eternal Word, to "become flesh" and die for us on the Cross. In this light, acts that manipulate, marginalize, or kill human persons in any phase or condition are grave offenses that should be proscribed by civil law in a civilized society.

6. Thomas Hobbes, *Leviathan* (London: Collier-Macmillan, 1962).
7. Immanuel Kant, *Groundwork of the Metaphysic Moments* (New York: Harper and Row, 1964).

Human Fulfillment

Finally, the Christian anthropological vision sees not only the intrinsic value of persons, but also their true fulfillment. Our being created in God's image reveals that we are most fulfilled by *freely giving ourselves to others*. While our capacities to reason and choose are evidence of our being made in the image of God, these capacities ultimately exist in service of our "social" or "transcendent" nature — our common vocation to freely share our gifts for the good of others. The relationship between freedom and fulfillment is particularly important for understanding Christian anthropology. Free human acts not only *reveal* the nature of "the self" to oneself, but also *constitute* the person, who thus cocreates him- or herself in a limited but real sense, and is therefore "autoteleological." Assisted by grace as a principle of action, we therefore become ourselves by freely giving ourselves to what is most valuable and capable of loving in return — other human persons and, ultimately, God. Conversely, we thwart our destiny when we refuse to enter into self-giving and self-sacrificing relationships oriented toward the good and the life of others. In contravening the dignity of others, we contravene our own human fulfillment. Human perfection, if not a gift from God, is self-destructive as an ideal.

According to this Christian anthropological vision, the human person who is created for communion with others is a unity of soul and body. In his catechetical reflections on human sexuality, the Pope reminds us that Aquinas' reflection on the doctrine of the resurrection of the body led to his definitive move away from a Platonic characterization of the real person as the immaterial soul, with the body as only a problematic prison. As better described by Aristotle, the soul is the animating or unifying principle or "form" of the body. The person is a unity comprised of a soul that expresses itself through and is *conditioned by* the body.

In a Christian vision of the person, the human body — including its sense faculties from which all knowledge of moral truths begins, and its *limitations* — is essential to the development and salvation of the whole person. God-given and natural bodily functionings, including sexuality, are normatively significant because they promote the perfection of the whole person. In his "personalistic" reading of the natural law tradition (in *Veritatis Splendor*), John Paul II has insisted that ethical principles derived by natural law from reflection on bodily finalities are not guilty of "physicalism," "naturalism," or "biologism."[8] Natural bodily functioning "constitute[s] a reference point for

8. See John Paul II, *Veritatis Splendor*, 46-50.

moral decisions" and "rational indications with regard to the order of morality." "Indeed, natural inclinations take on moral relevance only insofar as they refer to the human person and his authentic fulfillment." "The person, by the light of reason and the support of virtue, discovers in the body the anticipatory signs, the expression and promise of the gift of self, in conformity with the wise plan of the Creator."

In other words, both Christian revelation and reason tell us that the normally-functioning body which God provides "is good" because it promotes the perfection and the salvation of the whole person. The body, with its need for development and its limitations that temper our strength, intelligence, personality, and longevity, is integral to the development of virtues such as humility which allow us to give ourselves to others. Following St. Paul's cue, we also "rejoice in our weakness," our "treasure in earthen vessels," because it puts us in need of God, thus drawing us closer to the ultimate source of our fulfillment. Additionally, the complementarity of these limitations constitutes the "many gifts" that put us in need of, and draw us closer to, one another. What the Pope says in *Salvifici Doloris* about the salvific meaning of *suffering* also pertains to our *limitations*: both are "present in the world in order to release love, in order to give birth to works of love toward neighbor, in order to transform the whole of human civilization into a 'civilization of love.'"

The qualities that nontherapeutic genetic enhancements would seek to obtain can be good, and the motives for pursuing them may also be good. But eventually, one can imagine so-called "superhumans" who, less burdened by limitations, struggles, and the need for God and others, become self-centered and isolated, and thus less than fully human. The best life is therefore not one freed from all physical constraints. One could imagine this self-centeredness being directed against those who, lacking access to gene-enhancing technology, might find themselves part of a grossly disadvantaged and permanently objectified underclass of persons. Even those not directly benefiting from such enhancements, but effecting them, would find themselves in danger of seeing persons as products to be created according to their desires and needs, and of closing themselves off to the openness toward all that is the core of an authentic humanism.

The poem "Letter to Genetically Engineered Superhumans" by a young poet named Fred Dings nicely sums up many of these intuitions about genetic engineering that revelation supports:

> You are the children of our fantasies of form,
> our wish to carve a larger cave of light,

our dream to perfect the ladder of genes and climb
its rungs to the height of human possibility,
to a stellar efflorescence beyond all injury
and disease, with minds as bright as newborn suns
and bodies which leave our breathless mirrors stunned.
Forgive us if we failed to imagine your loneliness
in the midst of all that ordinary excellence,
if we failed to understand how much harder
it would be to build the bridge of love
between such splendid selves, to find the path
of humility among the labyrinth of your abilities,
to be refreshed without forgetfulness,
and weave community without the threads of need.
Forgive us if you must reinvent our flaws
because we failed to guess the simple fact
that the best lives must be less than perfect.[9]

A Christian vision of the future of bioethics must also emphasize the
need for the evangelization of *culture,* including the culture of universities
and the scientific communities. Because our nation does not include a common history, faith, race, or even language, law is often the most powerful car-
rier of culture in the United States. Major elements of our life together are
now regularly submitted for decision to the courts; and our culture's treatment of bioethical issues will inevitably be shaped by legal constraints and
the decisions of judges and lawyers regarding embryo experimentation and
storage, cloning, artificial conception procedures, prenatal diagnoses, abortion, genetic profiling, assisted suicide, and euthanasia and nontherapeutic
genetic engineering. It is more important than ever to work together to
achieve a Christian vision of bioethics and to consider carefully how best to
translate that vision and principles derived from it into our system of law.

9. Fred Dings, "Letter to Genetically Engineered Superhumans," from *Eulogy for a
Private Man* (Evanston: Northwestern University Press, 1999), p. 25. Reprinted by permission.

PART III

THE CHANGING FACE
OF HEALTH CARE

Money Matters in Health Care

SCOTT B. RAE

Being asked to address the area of money and health care is somewhat akin to the theologian who is asked to speak on "God, the Universe and Other Related Matters." It is a daunting task, with many unresolved issues. What follows is a discussion of the relationship between money and health care in the next few years and, in particular, the ethical issues that are raised.

The discussion of money and health care has been dominated recently by the phenomenon of managed care, how it has replaced traditional fee-for-service medicine, and the debate over the costs and benefits of such a system. Many physicians who enter practice are coming in under a variety of managed care payment plans, and they will know very little of standard fee-for-service medicine. This will increasingly be true for the next generation of health care professionals.

The Maturing of Managed Care

The contemporary health care marketplace is characterized by the maturing of managed care. That is, managed care is no longer new, and its benefits and shortcomings have become pretty clear, although the degree of managed care penetration in the health care market varies widely by geographic region. It has essentially delivered on its promise to control health care expenditures and increase access to care (though there are still too many people without access to quality care), but to the detriment of important aspects of health care such as patient choice, the patient-physician relationship and, at times, quality.

Indications of this maturing of managed care include:

1. Continued mergers of providers, namely hospital systems and physician groups, in order to keep their operating costs low enough to stay in business.
2. Some physicians moving away from HMO arrangements, settling for smaller practices, less income, and fewer headaches.
3. Some insurers and physician groups leaving the marketplace, either voluntarily or through financial pressure.
4. Formation of physicians' unions.
5. Physicians and hospitals successfully suing insurers over the denial of access to care.
6. A growing number of physicians supporting a single payer system.

At a fundamental level, the rise of managed care medicine, particularly for-profit managed care organizations, has raised a growing concern about the blend of business and medicine and about health care as a purchasable commodity. Although there is reluctant acceptance of managed care medicine by the general public, there is still widespread public skepticism about the ability of managed care organizations (MCOs) to protect the health interests of their customers — particularly when those interests conflict with the interests of their shareholders in maximizing their return on investment. Indeed, numerous bits of anecdotal evidence are now available that have not been encouraging and have added to the public skepticism about this new intersection of business and medicine. There are a growing number of disturbing stories in which patients have either been denied care or been given substandard care apparently for cost reasons, and have suffered or even died as a consequence.[1]

Business Ethics and Medical Ethics

It is widely assumed in this discussion that there are two different sets of ethical standards that are in fundamental conflict. Traditional medical ethics emphasizes patient well-being and autonomy and the physician's obligation to the patient. This stands in conflict with the way business ethics is widely perceived, with its emphasis on profit maximization. Many assume that these cannot be brought together and that managed care is an attempt, largely suc-

1. See for example, David R. Olmos, "Cutting Health Costs — or Corners?" *Los Angeles Times,* 5 May 1995.

cessful, to undercut traditional medical ethics with a business ethics that is only concerned with the bottom line. Numerous observers have stated that at some level, obligations to the bottom line will directly conflict with duties to serve the needs of patients.[2] An example of this assumed conflict between medical and business ethics can be found in a symposium on managed care that appeared in the well-respected journal, *Journal of Law, Medicine and Ethics.* Wendy K. Mariner puts it like this:

> The ethical principles that promote free and fair competition are quite different from the ethical principles that preserve the integrity of the physician-patient relationship and specifically those that protect patient welfare, and these principles can lead to quite different outcomes. MCOs were created to achieve economic objectives that may be fundamentally incompatible with traditional principles of medical ethics. Even if it is possible to agree that certain principles ought to apply to managed care, the market may make it impossible to live fully by those principles.[3]

She further argues that in the normal course of business an MCO can act as "ordinary business organizations with no moral obligations, or, at least, obligations that have little to do with traditional medical ethics."[4] She adds that when "an MCO's financial goals conflict with its service methods, little in the field of business ethics argues for giving subscribers priority."[5]

Much of the objection to the participation of shareholder-owned business organizations in managed care stems from concerns over the ethical challenges raised by the profit motive. Simply put, it is believed that there will be inevitable and irreconcilable conflicts in mixing business with medicine because the traditional "good" for medicine, dating back to the Hippocratic Oath, has been the health of the patient, while the "good" for business has been and continues to be profit.[6]

2. Arnold Relman, "What Market Values Are Doing to Medicine," *Atlantic Monthly* (March 1992): 106; and Bettijane Levine, "He Might Have the Cure for All Ills [Interview with AMA President Lonnie Bristow]," *Los Angeles Times,* 18 July 1995; and Joseph Cardinal Bernardin, "The Case for Non-Profit Health Care," speech delivered at the Harvard Business School Club of Chicago, 12 January 1995. Relman makes the pointed statement that "Medical care is in many ways uniquely unsuited to private enterprise. It cannot meet its responsibilities to society if it is dominated by business interests," p. 106.

3. Wendy Mariner, "Business Ethics vs. Medical Ethics: Conflicting Standards for Managed Care," *Journal of Law, Medicine and Ethics* 23 (1995): 236-246, at 236.

4. Mariner, "Business Ethics," p. 236.

5. Mariner, "Business Ethics," p. 238.

6. See William W. May, paper presented at the American Academy of Religion, panel

However, to suggest that business ethics is concerned only with profit maximization is to misunderstand what business ethics is about. Admittedly, the conduct of some HMOs around the country does reinforce the notion that all they are interested in is profit, and at any price. But a business ethics of profit maximization as the sole goal of business is an oxymoron, and some HMOs have been acting unethically in their pursuit of profit maximization.

It is a standard maxim in most business schools in the U.S. that the goal of a corporation is to maximize shareholder wealth. The classic statement of this is from Milton Friedman, who held that corporations were exercising all the social responsibility they needed to simply by making a profit. In doing so they provided the public with a useful product or service and provided jobs for people in the community. He further argued that for a corporation to act in a socially responsible way in ways that did not contribute to its profit was actually stealing from shareholders.[7] Friedman's paradigm has been challenged by stakeholder theory, which suggests that there are other stakeholders such as employees, the community, the environment, etc. whose interests should be taken into account along with those of shareholders in the decisions a company makes.[8]

A good example of this kind of socially responsible mixture of business and medicine is the pharmaceutical giant Merck.[9] A scientist at Merck discovered that a drug the company was selling to fight parasites in farm animals could be adapted to cure the Third World disease of river blindness. It, too, came from a parasite, one that grew to up to two feet long inside the human body and caused such suffering that victims frequently opted for suicide instead of enduring the agony. It eventually reached the eyes, causing blindness. The availability of the drug was great news to the victims, but the problem for Merck was that none of its potential "customers" could afford to pay for it. Merck was faced with a difficult decision: whether or not to invest the money necessary to develop the drug with no anticipation of any return on their in-

on "Managed Care: Insurers, Values, and the Bottom Line on Care," New Orleans, 25 November 1996.

7. Milton Friedman, "The Social Responsibility of Business Is to Increase Its Profits," *New York Times Magazine,* 13 September 1970, 33, 122-26.

8. Kenneth Goodpaster, "Business Ethics and Stakeholder Analysis," *Business Ethics Quarterly* 1:1 (January 1991): 53-73. See also Patricia Werhane, "Business Ethics, Stakeholder Theory, and the Ethics of Healthcare Organizations," *Cambridge Quarterly of Healthcare Ethics* 9 (2000): 169-81.

9. Cited in Andrew Wicks, "Albert Schweitzer or Ivan Boesky: Why We Should Reject the Dichotomy between Business and Medicine," *Journal of Business Ethics* 14 (1995): 339-51, at 345.

vestment. Merck had hoped to raise funding from a variety of public and private sources, but that effort predictably failed. The company essentially faced the decision of whether or not to give the drug away. It would cost them roughly $20 million annually to produce and deliver the drug, which they decided to do, forever. The reason they did so was not necessarily because they were doing charity, though they were. They were *mission driven.* Their corporate philosophy reads: "We try never to forget that medicine is for the people. It is not for the profits. The profits follow, and if we have remembered that, they have never failed to appear. The better we have remembered it, the larger they have been."[10] Their mission was not profit maximization at all costs; it was service for which they could expect a reasonable return on investment.

This outlook captures the spirit of Adam Smith.[11] Smith considered business a profession, in which service to the community through one's product/service was the mission, and profit was an anticipated by-product of excellence in that service. The service, not the profit, was the end. The bottom line was not the bottom line. His legacy for modern capitalism was not that which is widely attributed to him, that the sole goal of a corporation is to maximize shareholder wealth. In addition, for Smith, enlightened self-interest was the engine of capitalism. He never suggested, in the words of Gordon Gekko in *Wall Street,* that "greed is good." He distinguished between greed and self-interest, as does the Bible (Phil. 2:4, "look out not only for your own interests, but also for the interests of others").

More importantly, Smith assumed that individuals possessed the internal resources necessary to show restraint in their pursuit of self-interest. That is, ethics and self-restraint play critical roles in the proper functioning of the market and everyday market transactions. He was very sensitive to this, out of his profession as a moral philosopher, and he assumed that the Judeo-Christian moral consensus that existed in his time would function as the source of this restraint. Of course, that consensus is not what it used to be. But restraint of self-interest is still necessary for a properly functioning economic system. As Catholic theologian Michael Novak argues, "A firm committed to greed unleashes social forces that will sooner or later destroy it.

10. Quoted in Wicks, p. 345.

11. For further discussion on the contribution of Smith to business ethics, see Patricia Werhane, *Adam Smith and His Legacy for Modern Capitalism* (New York: Oxford University Press, 1991). For a specific application of Smith to the business of medicine in managed care see, "The Ethics of Health Care as a Business," *Business and Professional Ethics Journal* 9:3-4 (Fall-Winter, 1990): 15. See also the fine discussion of a business ethics applied to health care in Kenman L. Wong, *Patients and Profits: An Ethic for Managed Care* (Notre Dame: University of Notre Dame Press, 1999).

Spasms of greed will corrupt its executives, anger its patrons, injure the morale of its workers, antagonize its suppliers and purchasers, embolden its competitors and attract public retribution."[12] Though it is true that some firms operate on the "greed is good" corporate philosophy, to characterize "business ethics" as such is both inaccurate descriptively and also an internal contradiction.

This stakeholder approach to business ethics is also consistent with the way some of the most successful companies conduct their business. In their best-seller, Stanford University management professors James Collins and Jerry Porras describe what they call "visionary companies," that is, companies that are "built to last."[13] What sets these companies apart, besides their long-term financial profitability, is the set of core values that governs their day-to-day business. Collins and Porras summarize their findings as follows:

> Contrary to business school doctrine, "maximizing shareholder wealth" or profit maximization has not been the dominant driving force or primary objective through the history of the visionary companies. Visionary companies pursue a cluster of objectives, of which making money is only one — and not necessarily the primary one. Yes, they seek profits, but they are equally guided by a core ideology — core values and a sense of purpose beyond just making money. Yet paradoxically, the visionary companies make more money than the more purely profit-driven companies.[14]

In other words, these companies operate under more of a stakeholder approach to their mission. They balance a host of competing obligations, including those to shareholders, customers, suppliers, the community, and the environment. As applied to health care, this model is a great improvement over the perceived view of business ethics. Clearly, the interests of patients can coexist with the interests of shareholders. Business and health care have consistently mixed in the past century and will continue to do so. It will not do to suggest that they cannot or should not mix, and to think that the difficulties will disappear. The task of ethics is to insure that the interests of patients are not compromised in the pursuit of profit.

---

12. Michael Novak, *The Spirit of Democratic Capitalism* (Lanham, MD: University Press of America, 1991), 92.

13. James Collins and Jerry Porras, *Built to Last* (New York: Harper Business, 1994).

14. Ibid., p. 8. Cited in Werhane, "Business Ethics, Stakeholder Theory," p. 172.

Future Trends and Ethical Challenges

While attempting to foresee the future fully is folly, there are some areas in which money and health care will predictably continue to intermingle, with significant ethical issues raised. Some of these are more macro issues that will be resolved only through broad changes to the health care delivery system. These are issues that deal with distributive justice, i.e., the philosophical and theological debate over how the goods of society should be distributed. Health care delivery is one of the most recent social debates that has questions of distributive justice at its core. Questions concerning the criteria by which health care is distributed — need, merit, social worth, ability to pay, or some combination of these — are at the heart of the current discussion of health care reform. The door is open for theologians and philosophers to bring distributive justice reflections into the debate.

1. Ranking Priorities in Health Care Delivery

The current health care delivery system is attempting the impossible. It is trying to achieve four goals simultaneously that cannot all be achieved due to a scarcity of resources to devote to health care.[15] These four goals are universal access, high quality of care, cost control, and freedom of patient choice. At present, the health care system under managed care is accomplishing the goals to this degree: access is increasing, but there is still a long way to go; quality of care is generally good and patients are quite satisfied, until one becomes seriously ill; cost control has improved (managed care appears to have delivered on its promise to control costs), though it appears that managed care has now squeezed most of the waste and fat out of the system (witness the most recent rounds of premium increases in the past two years); and choice is significantly limited.

We first need to admit that the supply of resources available for health care is insufficient to meet the need because it is limited by other priorities. In the 1980s and early 1990s there was consensus that health care costs were spiraling out of control, and that if they were not contained, they would constitute a disproportionate share of society's resources. This spiraling was due to a variety of factors, including overuse of expensive high-tech treatments, especially at the end of life, the legacy of aggressive treatment, and the phenom-

15. H. Tristram Englehardt, *Foundations of Bioethics,* 2nd ed. (New York: Oxford University Press, 1996), pp. 375-410.

enon of third-party payers removing patients from payment decisions. Managed care emerged as a solution to cost containment problems, and as such, has likely run its course. There is some movement away from capitation as a mechanism to control costs; HMO enrollments have flattened out in recent years, as have the senior HMO enrollments; and the price difference in premiums between PPO and HMO arrangements is narrowing. We are seeing more growth in PPO arrangements that allow for greater choice, on behalf of both patients and physicians, though capitation is likely here to stay as long as health care is distributed on a market model.

Secondly, we need to address the issue of whether health care is a right or a commodity. The more that market mechanisms are used to control costs and insure quality and access, the more health care is regarded as a commodity that can be purchased and sold in the market like any other product or service. Of course, health care is unlike most other services, due to the immense knowledge differential between patients and physicians and the urgency of securing health care when one is sick. However, there is some parallel between health care and legal representation, which is generally considered a market service (except for public defenders). Other similar goods and services involve necessities such as housing, food, and education. Essential goods and services are generally distributed according to the market and one's ability to pay, unless one cannot afford them — at which point a minimal standard is provided through a vehicle such as public housing, food stamps, or Medicaid. There is a clear difference in quality between the two tiers in the current health care arrangement, and even the lower tier of minimum "guaranteed" health care still has access problems and issues of cost control. What does seem to be unique about the way health care is regarded is that people are increasingly asserting a positive right to all the health care they need, regardless of their ability to pay. Health care is perhaps the only item on the market today where people expect all they need without having to worry about paying for it.

2. Balancing the Interests of Individual Patients and the Patient Population Being Served

Traditional medical ethics has put the patient's interests ahead of everything else, irrespective of the costs involved. This approach has assumed that resources were not an issue and that there were no competing interests other than those of the individual patient. But with managed care bringing scarcity of resources to the forefront of society's attention, the physician's primary ethical obligations became divided. With payment schemes such as capitation in place, there was the prospect of conflict between the patient's best medical interest and the physician's/hospital's financial interest. Capitation eliminated vehicles like cost shifting that had been used to help pay for care for the uninsured, and the scarcity of resources became more visible. The ethics of this situation seems clear, that the fiduciary relationship between patient and physician demands that the physician put the patient's interest ahead of his or her own self-interest. Further it is not unreasonable for physicians to disclose when there might be a conflict of interest. For example, some physicians disclose to all patients that they have a financial interest in the lab to which they refer all their lab work. Others would even have physicians disclose to patients the way in which they are paid for patient care.

However, the more difficult conflict raised by scarcity is the conflict between the interests of individual patients and the interests of the entire patient population served by the physician/hospital.[16]

With a fixed amount of resources available to treat a patient population, there is a possible conflict between those interests. The conflict is particularly acute under a payment system such as capitation, in which the amount of resources available to treat a given population is fixed. Here physicians have a contractual obligation to care for the entire patient population for which they have contracted, and patients who utilize a disproportionate share of those resources jeopardize the ability of physicians to provide a proper level of care for all their patients. Of course, the degree to which the physician is compensated by capitation will determine how acute this conflict will be. There is less of a conflict when payment is made under a PPO arrangement, though physicians do have to see a large number of patients and deal with many administrative details in order to maintain their incomes.

3. Saying "No" to Families Requesting
Futile Treatment at the End of Life

Virtually every day in hospitals and medical centers around the country, families are making inappropriate requests for aggressive treatment at the end of life. More often than not, physicians accommodate these requests out of a fear of being sued or to avoid the tension in dealing with the family. The result is

16. This is developed in more detail in E. Haavi Morreim, *Balancing Act: The New Medical Ethics of Medicine's New Economics* (Georgetown: Georgetown University Press), 1995.

that resources are unnecessarily spent at the end of life on futile or burdensome treatments for patients where the cost is very high and the benefit is minimal. In the future, physicians and hospitals will more often say "no" to these kinds of requests for the right reason: a proper stewardship of scarce medical resources. To be sure, there will be cases in which treatments should have continued, but in the vast majority of them, there is medical consensus that aggressive treatment should be stopped and a regimen of palliative care be initiated, perhaps including transfer to hospice care.

An example of how some states are addressing this problem is in California, where a new law (AB 891) gives physicians and hospitals the right, and immunity from liability, to say no to requests for any treatment (particularly at the end of life) which is contrary to a hospital's policy, violates the standard of care, or violates the dictates of the physician's or hospital's conscience. This will empower physicians to refuse unreasonable requests for end-of-life aggressive treatment as long as the hospital has a futility policy and the refusal of treatment is in accord with community standards of care. The law reads: "A health care provider may decline to comply with an individual health care instruction or health care decision if the instruction or decision is contrary to a policy of the institution that is expressly based on reasons of conscience and if the policy was timely communicated to the patient. . . . A health care provider or health care institution may decline to comply with an individual health care instruction or decision that requires medically ineffective health care or health care contrary to generally accepted health care standards applicable to the health care provider or institution."[17]

Such legislation is the coming trend and will enable hospitals to become better stewards of scarce resources. It is unethical for families to demand futile treatment, or particularly to demand treatment that is more burdensome than beneficial for their loved ones. In some cases, such demands are tantamount to torture of loved ones in the name of "doing everything." Families need help to envision more accurately what the treatments they authorize will do to their loved ones. Moreover, once treatment is under way, they need to see what is being done to their loved ones so that they do not inappropriately persist in demanding futile or burdensome treatment at the end of life.

One way in which patients can help their family members is to refuse futile or burdensome treatment at the end of life. We should consider it virtuous for a patient who wishes to refuse a treatment at the end of life because

17. AB 891, California, sections 4734-4735. See also Susan M. Wolf, "Health Care Reform and the Future of Physician Ethics," *Hastings Center Report* 24 (March-April 1994): 28-40.

another use of the same resources would be more justified ethically. It is odd that only with illness does the patient acquire the right to have his/her interests be the only interests to be considered.[18] Normally, in families, individual members weigh the interests of other family members with their own interests. It is rare that any family members would require or expect that their interests are the only ones that matter. In fact, we teach our children that the world does not revolve around them, and that in a family, there are competing interests that must be weighed carefully. Why should this be any different at times of illness? To be sure, the vulnerability of ill people means that their interests may be much greater than those of others and may need to be more carefully safeguarded. However, these considerations should not exempt them from weighing their interests against those of the rest of the family. Where continuing aggressive treatment would constitute a crushing — even potentially life-threatening — burden to one's family, we should consider it virtuous to refuse such treatments. To say that there is a duty to refuse such treatments is going too far, since such a duty would constitute a claim on the ill person that could be exercised by one's loved ones. This claim has great potential to be coercive, and should be rejected. But it would not be too much to commend as virtuous patients' rejection of inappropriate treatment.

4. Ethics Committees Giving Increased Attention to Organizational Ethics

In the last five years, ethics committees have been asked to shift their focus from strictly clinical ethics to the business side of health care, or organizational ethics. Issues include billing, marketing, admission/discharge/transfer policies, and joint ventures and affiliations with other health care institutions. The Joint Commission which accredits health care organizations has shown much greater interest recently in this part of the ethics committee's work. Increasingly, the ethics focus in health care will be on organizational matters, where money and morality mix more frequently and more intentionally. Those who are involved in ethical reflection in hospitals and medical centers will be asked to blend business ethics and medical ethics, to formulate an ethics that will serve both patients and organizations. It would strengthen the bioethics community to become better educated in the sister field of business ethics to better serve health care institutions.

18. John Hardwig, *Is There a Duty to Die?* (New York: Oxford University Press, 1998).

5. Gaming the System

Though there have always been temptations to manipulate the system for one's personal gain, under managed care there have been opportunities to work around the system in order to insure, not only the most profitable reimbursement, but also the patient's best interest. For some physicians and hospitals, upcoding is a more time-efficient way to secure that which they believe is medically necessary for their patients.[19] Some physicians are convinced that the only way to maintain their viability in practice is to "game" the system regularly because, as it is currently constituted, they do not receive adequate reimbursement to stay in business. The alternative to upcoding is to spend countless hours advocating for patients, either by phone or by letter. Case managers in hospitals are tempted to exaggerate the accounts of a patient's illness and course of treatment and falsify diagnoses in order to obtain reimbursement or authorization for further treatment. This has become problematic particularly under managed care, which has tightened the requirements on physicians for authorizing treatments and for reimbursing procedures due to the emphasis on controlling health care expenditures. Changes in the system will present new opportunities to circumvent further attempts at cost control.

Gaming the system is morally problematic precisely because of the deception it involves. There is a subtle difference between writing the narrative of a patient's course of disease and treatment dramatically and falsifying the diagnosis in order to receive authorization or reimbursement. Such falsifying is deception and should not be done. Under the law, it constitutes insurance fraud. Numerous physicians have given anecdotal evidence that gaming the system is not necessary to achieve a good outcome for one's patient. It does save time and headaches, but if a physician is willing to go through the proper channels, it does usually pay off, thus making gaming the system unnecessary. A further problem with gaming the system is that it is fundamentally discriminatory, that is, one must be selective about which patients for whom one games the system. For if a physician did this for all patients for whom he/she deemed it necessary, it would be detected.

19. Victor E. Freeman, et al., "Lying for Patients: Physician Deception of Third Party Payers," *Archives of Internal Medicine* 159 (October 25, 1999): 2263-70; Warren Kinghorn, ed., "Intentional Deception by Physicians," *Journal of the American Medical Association* 282:17 (November 3, 1998): 1674-79.

6. There Must Be a Christian Concern for the Poor and Those without Access to the Health Care System

Many of the most difficult issues in medical ethics have to do with access to care for the uninsured and underinsured. Though it is true that some are uninsured, or underinsured, by choice, or temporarily while in between jobs which provide coverage, the majority of these feel vulnerable and would be very vulnerable if hit with a serious illness. This is particularly the case in states with a large immigrant population, such as Texas and California, in which the immigrants come from cultures that tend to view health care as a right. Vulnerability will also extend to the rapidly swelling ranks of the elderly who qualify for Medicare, as the baby boomers reach retirement age, with a declining work force available to provide support.

Regardless of the way in which the macro issues are resolved, an integral part of the health care system, particularly if the trend toward the market continues, will be the provision of care for the poor from charitable and religious groups. Market forces should not be allowed to force physicians and hospitals away from providing a degree of charity care. Especially for Christian physicians, who follow in the tradition of the healing ministry of Christ, the biblical mandate to care for the poor is a powerful one. This is why my hope is that not-for-profit health care continues to be financially viable, because I fear for the poor if all health care becomes dominated by the for-profit sector. Christ's admonition to care for the vulnerable surely applies to the poor when they hit that most vulnerable state of illness and decline in their health.

Spirituality and Alternative Medicine

DÓNAL P. O'MATHÚNA

Alternative medicine continues to grow in popularity. Estimates are that approximately thirty percent of Americans used some form of alternative medicine in 1990, and by May 2000, this had increased to thirty-five percent of more than 46,000 adults surveyed.[1] While it continues to be plagued by unclear definitions, alternative medicine also raises new ethical issues. One of these concerns the role of health care professionals in the provision of "spirituality." Part of alternative medicine's appeal is its claim to satisfy people's hunger for spirituality and the transcendent. Some proponents, like Deepak Chopra and Andrew Weil, act more like religious prophets delivering inspirational sermons than physicians offering evidence-based recommendations.

Americans are interested in spirituality, as evidenced by surveys and statistics. For example: 75 percent of hospitalized patients want physicians to consider their spiritual needs.[2] More than 90 percent of Americans pray, and 95 percent claim their prayers are answered.[3] More than three-quarters of all Americans believe God answers prayer for healing an incurable illness, and 14 percent claim they have experienced such healing.[4] One survey has found that 99 percent of family physicians believe religious belief can contribute to

1. "The Mainstreaming of Alternative Medicine," *Consumer Reports* 65:5 (May 2000): 17-25.

2. Richard P. Sloan, et al., "Should Physicians Prescribe Religious Activities?" *New England Journal of Medicine* 342:25 (June 2000): 1913-6.

3. Scott R. Walker, et al., "Intercessory Prayer in the Treatment of Alcohol Abuse and Dependence: A Pilot Investigation," *Alternative Therapies in Health and Medicine* 3:6 (November 1997): 79-86.

4. Walker, "Intercessory Prayer," pp. 79-86.

patients' healing; 92 percent of HMO professionals believe likewise.[5] About two-thirds of these same physicians and HMO professionals say they use prayer or meditation themselves when they are ill.[6]

This interest has opened doors for spiritual discussions with patients that would not have been possible a few years ago. A bridge has been built over the previously-dug gulf between medicine's focus on physical dimensions, and patients' emotional, relational, and spiritual concerns. "Holistic care" was once a taboo term; now it raises fears only in the minds of hardened skeptics. In fact, the Bible promotes what can only be described as holistic care.[7] God wants us to love him in body, mind, and spirit, and provides the resources to promote the health of all three through the abundant life he offers (cf. John 10:10).

Another recent trend that is becoming a big factor in this millennium is the increased emphasis placed on evidence-based health care decisions. Whether those paying for health care call it "outcomes assessment," "quality assurance," or "clinical pathways," they insist that those providing therapies justify their clinical decisions. This trend has arisen for more than financial reasons, for it is rooted in the desire of its proponents to provide the safest, most effective therapies available. A good argument can be made for evidence-based decision making on the basis of the need for stewardship, especially in an age of limited resources.

Although alternative therapies often have little more than anecdotal support, medicine should be open to those therapies and herbs that demonstrate clinical effectiveness and safety in controlled trials. Unfortunately, as research demonstrates that some herbs are effective for specific conditions, like St. John's wort for mild depression and anxiety, evidence of side effects and drug interactions is also being revealed.[8] Similarly, the British Medical Association reported that while evidence supports the use of acupuncture for nausea and vomiting, back pain, dental pain, and migraine, the evidence is

5. Yankelovich Partners, *Belief and Healing: HMO Professionals & Family Physicians* (August 1997), 12.

6. Yankelovich Partners, *Belief and Healing*, p. 28.

7. Dónal P. O'Mathúna and Walt Larimore, *Alternative Medicine: The Christian Handbook* (Grand Rapids: Zondervan, 2001); Dónal P. O'Mathúna, "Emerging Alternative Therapies," in *The Changing Face of Health Care: A Christian Appraisal,* ed. John F. Kilner, Robert D. Orr, and Judith Allen Shelly (Grand Rapids: Eerdmans, 1998), pp. 258-79; Judith Allen Shelly and Arlene B. Miller, *Called to Care: A Christian Theology of Nursing* (Downers Grove, IL: InterVarsity, 1999).

8. E. Ernst, "Second Thoughts About Safety of St John's Wort," *Lancet* 354 (December 1999): 2014-6.

equally clear that it is ineffective for asthma, smoking cessation, weight loss, recovery from stroke, rheumatic diseases, and tinnitus.[9] It is becoming harder to claim that a particular therapy works for everything and anything, or is completely harmless. This millennium is bringing with it an end to the free ride that alternative medicine has enjoyed until now.[10]

Spirituality has not escaped this concern for scientific validation. The impact of spirituality or religion on health and healing has been investigated by modern research methods. Again, while the renewed attention can be welcomed, Christians should approach this area carefully. If we discover that those who attend church less often are less healthy, should church attendance be prescribed? If so, which church, which denomination, which religion? If we discover that patients who are prayed for do not recover any faster, should we therefore not pray for patients? What if a study finds that the death rate among those receiving more prayer is higher? Could a doctor be sued for negligence for failing to caution family members of the risks of praying? If prayer works, what do we say to the person who converts in order to obtain God's healing and finds he is about to die anyway?

These questions raise some of the core issues requiring attention before health care professionals become providers of spirituality. Christians are called to test the spirits and see if they are from God (1 John 4:1-2). Christian health care professionals should therefore be directly involved in examining what is being provided as "spirituality." This is an opportunity to make our faith more apparent to others. But we must also be willing to address the dangers inherent in reuniting medicine and religion. In addition, the fact that so many are turning to health care professionals to meet their spiritual needs should be a call for serious reflection on the part of today's churches.

The Connection between Religion and Health

Numerous studies have found a positive connection between religious commitment and health status.[11] For example, those scoring high on religious

---

9. British Medical Association, *Acupuncture: Efficacy, Safety and Practice* (Amsterdam: Harwood Academic, 2000). See also, NIH Consensus Development Panel on Acupuncture, "Acupuncture," *Journal of the American Medical Association* 280:17 (November 1998): 1518-24.

10. Marcia Angell and Jerome P. Kassirer, "Alternative Medicine — The Risks of Untested and Unregulated Remedies," *New England Journal of Medicine* 339:12 (September 1998): 839-41.

11. Dale A. Matthews, et al., "Religious Commitment and Health Status: A Review

commitment have reported less depression, lower rates of suicide, less abuse of alcohol and other drugs, lower rates of cancer and heart disease, and lower mortality rates. In one study, women treated for hip fractures who have held to their religious beliefs strongly have been less depressed and better able to walk further upon discharge than those who rated themselves as less religious.

While these results are encouraging, we must be cautious. Reviewers of this literature point out that the measures used to evaluate religious commitment have often been simplistic and external. For example, church attendance or church membership has frequently been used. This approach may actually promote this sort of external religion, especially when the headlines read something like "Church Attendance Boosts Health" or "Spirituality Increases Survival Rate."

In many cases, people self-report their spiritual activities or level of involvement by thinking back over the previous week or month. We have a tendency to report doing better than we actually did on activities we know we should have been doing, so these types of surveys can give unreliable results. There are also several studies which have found conflicting results, and others which have found negative results.[12]

We cannot forget (or excuse) the fact that much harm has been done in the name of religion. Physical health can be harmed by certain religious beliefs and practices. In the view of Christian Science, illness and pain are illusions, and medical treatment is rejected. This type of outlook is part of much New Age spirituality, which will likely lead to more cases of people refusing standard medical treatment for religious reasons.

One study has found that between 1975 and 1995 in the U.S., 172 children died who did not receive standard medical care because their families belonged to various religious sects opposed to the use of conventional medicine.[13] More than eighty percent of these children would have had better than a ninety percent chance of survival if they had received standard preventive medical care or treatment. The parents of these children, and the leaders of their groups, sincerely believed they were doing the right thing in God's eyes. This area of research is complex and new, and we must therefore be cautious and tentative as we make recommendations based on these studies.

---

of the Research and Implications for Family Medicine," *Archives of Family Medicine* 7:2 (March-April 1998): 118-24.

12. Jeffrey S. Levin and Harold Y. Vanderpool, "Is Frequent Religious Attendance *Really* Conducive to Better Health? Toward an Epidemiology of Religion," *Social Science & Medicine* 24:7 (1987): 589-600.

13. Seth M. Asser and Rita Swan, "Child Fatalities from Religion-Motivated Medical Neglect," *Pediatrics* 101:4 (April 1998): 625-9.

Current research also seems to support the health benefits of any form of spirituality. Many today want to define spirituality as something very different from biblical spirituality. One proposed nursing curriculum defines spirituality as "the unifying force of a person; the essence of being that shapes, gives meaning to, and is aware of one's self-becoming. . . . It is expressed and experienced uniquely by each individual through and within connections to God/Life Force/the Absolute, the environment, nature, other people, and the self."[14] Herbert Benson states that although he believes in God, "People don't have to have a professed belief in God to reap the psychological and physical rewards of the faith factor."[15] For him, faith is just another therapy that brings medical benefits: "faith is a natural and inevitable physiologic reaction to the threats to mortality we all face."[16] This view proclaims that having faith is what matters, not the content of one's faith.

Approaching faith for its health benefits is, to paraphrase C. S. Lewis, putting second things first. We can expect that, in general, faith in God will have spiritual, physical, emotional, and relational benefits. But belief in God should be grounded in faith and trust in him, not health benefits. The second things cannot withstand being put first in our lives. We will become frustrated if we view involvement with God as first and foremost a way to develop friendships, or to have some great experiences, or to gain health. These are potential benefits, but should never be our primary desire or purpose for our faith. As Lewis reiterates, "you can't get second things by putting them first; you can get second things only by putting first things first."[17]

Jesus occasionally took care of his disciples' needs by miraculously feeding thousands with a little bread and a few fish (John 6:1-14). But the very next day he refused to feed them miraculously, replying to their pleas, "I tell you the truth, you are looking for me, not because you saw miraculous signs but because you ate the loaves and had your fill" (John 6:26). Faith's "first thing" is coming into relationship with God, not good health.

In addition, we must be careful not to go beyond what this data shows, because it is not clear from the studies that faith in Jesus Christ is what brings better health. Dale Matthews has reviewed much of this literature in his popu-

14. Barbara M. Dossey, ed., *Core Curriculum for Holistic Nursing,* American Holistic Nurses Association (Gaithersburg, MD: Aspen, 1997), p. 42.

15. Herbert Benson, *Timeless Healing: The Power and Biology of Belief* (New York: Scribner, 1996), p. 156.

16. Benson, *Timeless Healing,* p. 300.

17. C. S. Lewis, *God in the Dock: Essays on Theology and Ethics,* ed. Walter Hooper (Grand Rapids, MI: Eerdmans, 1970), pp. 278-81.

lar book, *The Faith Factor*.[18] He describes twelve "remedies" integral to religion that have important health benefits, including development of self-control, community, trust, transcendence, and love. But he acknowledges that all these factors can be found in nonreligious settings. For him, this research "reveals that religious involvement can be seen as a unique 'combination agent' that efficiently delivers a series of powerful, interrelated ingredients promoting health and well-being."[19] There are other ways to get these benefits, but the best way according to Matthews is through "religious organizations."

This finding should not surprise us. When we live according to God's design, however incompletely, we will be healthier. God designed us for intimate fellowship with others, so when people have good friends — whether within their families, neighborhoods, clubs, or churches — they will reap the benefits of living more as God designed them. But we go too far when we preach a health and wealth gospel, one that purports that those who follow God's ways will live long and healthy lives. This may happen, but it is not guaranteed. The Scriptures describe a number of godly men and women who got sick. Epaphroditus became sick because of his diligence to Christian work and almost died (Phil. 2:25-30). The Bible does not guarantee health, and in fact promises that all Christians will suffer (Matt. 24:4-22; 2 Cor. 1:3-11).

When spirituality and health are linked, health wrongly becomes the measure of one's spirituality. Those who are sick are seen as having insufficient faith to produce healing. This can inflict much guilt on those who are sick, especially if they become terminally ill. Yet the Bible maintains that times of suffering and pain can result in spiritual growth. "Consider it pure joy, my brothers, whenever you face trials of many kinds, because you know that the testing of your faith develops perseverance. Perseverance must finish its work so that you may be mature and complete, not lacking anything" (James 1:2-4). Christians should not expect freedom from suffering because of their faith, but should expect additional suffering because of their faith (1 Peter 4:12-19).

Turning to spirituality for the sake of one's health quickly leads to a form of religion which is self-focused and self-absorbed. One's own physical and emotional health becomes more important than anything else. In contrast, Jesus proclaimed, "It is more blessed to give than to receive" (Acts 20:35). We should care for our health as a matter of stewardship, but there are times when it is appropriate to place our physical health in danger, for the

18. Dale A. Matthews, *The Faith Factor: Proof of the Healing Power of Prayer* (New York: Viking, 1998), pp. 42-52.

19. Matthews, *The Faith Factor*, p. 41.

higher goal of worshiping God or serving others. "What good will it be for a man if he gains the whole world, yet forfeits his soul? Or what can a man give in exchange for his soul?" (Matt. 16:26). We may even be called to give up our physical life for the good of others, since "Greater love has no one than this, that he lay down his life for his friends" (John 15:13).

We too easily turn our health into an idol, fall down, and worship it. When we tie our health to spirituality, implying that good health is evidence of spiritual growth, we increase the temptation to make the pursuit of health the purpose of our lives. Those who worship the true God know that this should not be the case. Therefore we praise missionaries who go to foreign lands where they, and their families, will not have access to all the medical care they would have at home. We praise those doctors, nurses, chaplains, and volunteers who hold the hands and wipe the sweat off those who are the lepers of our day. And when faced with health care decisions that involve unjust allocation of resources, cutting-edge technology that is ethically questionable, or alternative therapies that are spiritually questionable, we should be willing to say "No!" Attempts to prove that spirituality leads to physical health can make these sacrifices harder. So the evidence connecting spirituality and physical health is helpful to the extent that it provides support for allowing Christian health care professionals to encourage patients to trust in Christ. However, as we have seen, there are many ways that this connection can be misunderstood or misapplied.

Intercessory Prayer for Healing

Interest in spirituality has led to research on the effectiveness of intercessory prayer. Harris and colleagues reported a major study in 1999 that led to nine pages of letters and a four-page commentary a few months later in the journal. This demonstrates how controversial and topical this issue has become. But this interest is not new. In 1872, John Tyndall, responding to recent scientific developments in England, proposed that all Christians pray for patients in one London hospital for at least three years.[20] He predicted that the death rate for patients at that hospital with a particular disease would be no different than the death rates at other hospitals, which would confirm his skepticism of Christianity. Tyndall's proposal, like the Harris study, generated a storm of controversy, reproduced in the fascinating book, *The Prayer-Gauge*

20. Stephen G. Brush, "The Prayer Test," *American Scientist* 62 (September-October 1974): 561-63.

Debate.[21] Francis Galton, a cousin of Charles Darwin, contributed a proposal that since people pray specifically and frequently that clergy and royalty live long and healthy lives, their life expectancies would reveal whether or not prayer works. Galton found that clergy and royalty had shorter life expectancies than other professionals who were regarded as worldlier and less prayerful (lawyers and doctors). Thus began the scientific controversy that continues today.

Probably the best-known study was published in 1988 by Randolph Byrd.[22] This Christian cardiologist carried out a double-blind, randomized study with 393 patients in the intensive care unit at San Francisco General Hospital. The group that "born again" Christians prayed for did significantly better in six ways than those who received no additional prayer through the research. However, Byrd himself admitted that these six individual results "could not be considered statistically significant because of the large number of variables examined."[23] Byrd measured twenty-three other outcomes which showed no statistically significant differences between the prayed-for and control groups. The former received specific prayer for a rapid recovery and the prevention of death, but the two groups did not differ in these outcomes.

Byrd developed an assessment tool to give an overall score for how well the patients did during their hospitalization. This assessment tool was not "validated," which is an important way to ensure that the assumptions and calculations involved in the tool are accurate and reliable. It was this tool that led to the conclusion that patients in the prayer group did significantly better overall than those in the control group.

The study by Harris and colleagues repeated the Byrd study, this time with 990 patients admitted to the coronary care unit in Kansas City, Missouri.[24] The patients were randomized into two groups without knowing they were in a study, which raises serious ethical questions. We as Christians may see little problem with praying for someone without asking their permission. However, consider your reaction to being included in a study where spirit guides were sent to your assistance without the researchers informing

21. Prof. Tyndall, Francis Galton, et al., *The Prayer-Gauge Debate* (Boston: Congregational Publishing Society, 1876); Galton's article excerpted as, "Does Prayer Preserve?" *Archives of Internal Medicine* 125 (April 1970): 580-1, 587.

22. Randolph C. Byrd, "Positive Therapeutic Effects of Intercessory Prayer in a Coronary Care Unit Population," *Southern Medical Journal* 81:7 (July 1988): 826-9.

23. Byrd, "Positive Therapeutic Effects of Intercessory Prayer," p. 829.

24. William S. Harris, et al., "A Randomized, Controlled Trial of the Effects of Remote, Intercessory Prayer on Outcomes in Patients Admitted to the Coronary Care Unit," *Archives of Internal Medicine* 159 (October 1999): 2273-8.

you of this. Now picture this happening as you lie in an intensive care unit. The highest standards of research ethics should be called for, and adhered to, when investigating spiritual issues, and these require informed consent for participation.

The Harris study used clusters of five Christians to pray for one of the study groups for twenty-eight days. Thirty-five different medical outcomes were measured to see how the two groups differed. No statistically significant differences were found for any individual outcome. The researchers used two tools to generate two overall scores. The tool developed by Harris and colleagues showed no significant differences between the two groups. Byrd developed his own tool, but did not validate it. This tool showed that the prayer group scored an average eleven percent better than the control, which was statistically significant. However, the two studies together give unclear and conflicting results, which points to the difficulties and complexities involved in this type of research.

When this chapter was being prepared, an exhaustive search of the literature revealed fifteen studies where the effects of intercessory prayer on patients' health were measured in controlled environments. (The several additional prayer studies that have been published while this book was in press only confirm the pattern of results found by the other studies.) Apart from the two studies just described, the other studies have all been different, using different types of prayer for patients with many different health conditions. For the purpose of briefly describing these studies, three broad groupings will be used: (1) higher-quality studies of physical conditions, (2) small studies examining physical conditions, and (3) studies of psychological conditions.

The first group of studies, including the Byrd and Harris studies, includes four higher-quality studies with larger groups of subjects. A pilot study called the MANTRA Project had 150 cardiac patients, but was not designed to give statistical evaluations.[25] Patients received various forms of Christian prayer, and also Jewish, New Age, and Buddhist prayer for six months. The researchers reported encouraging results and are currently conducting a larger study. The fourth study in this group involved forty patients with AIDS.[26] Half the patients received prayer from Christian, Jewish, Buddhist, Native American, shamanic, bioenergetic, and meditative healers. At

25. Bonnie Horrigan, "Mitchell W. Krucoff, MD: The MANTRA Study Project," *Alternative Therapies in Health and Medicine* 5:3 (May 1999): 75-82.

26. Fred Sicher, et al., "A Randomized Double-Blind Study of the Effect of Distant Healing in a Population with Advanced AIDS: Report of a Small Scale Study," *Western Journal of Medicine* 169:6 (December 1998): 356-63.

the end of this period, eleven outcomes were measured, and the people in the prayer group did significantly better in six.

The second group of studies contains six small studies (with between sixteen and fifty-three subjects) examining the effects of intercessory prayer on conditions like diabetes, blood pressure, pain after hernia surgery, and leukemia. Three used Christian prayer, and three used a variety of energy-directing forms of prayer. Only two showed that those who were prayed for did significantly better than those who did not receive the additional prayer. One of these involved Christians praying for children with leukemia.[27] The other found that people reported less pain after having their molar teeth extracted when they also received Reiki and LeShan prayer.[28] The latter is a natural (as opposed to supernatural) form of prayer that allegedly allows one person to stimulate another person's natural capacity for self-healing. The four other studies did not reveal significant benefits from prayer.[29]

The third group of studies examines psychological measurements and includes five studies. The two earliest ones were not rigorously designed in that the patients were not assigned to groups randomly or in a blinded manner.[30] Thus, although those who were prayed for demonstrated some improvements, factors other than prayer could account for those results. Three studies were published during the 1990s and were randomized and double-blinded. One used LeShan prayer for patients with major depression and measured changes in their levels of depression and well-being.[31] The second

27. Platon J. Collipp, "The Efficacy of Prayer: A Triple-blind Study," *Medical Times* 97:5 (May 1969): 201-4.

28. D. P. Wirth, et al., "The Effect of Complementary Healing Therapy on Postoperative Pain after Surgical Removal of Impacted Third Molar Teeth," *Complementary Therapies in Medicine* 1:3 (July 1993): 133-8.

29. C. R. B. Joyce and R. M. C. Welldon, "The Objective Efficacy of Prayer: A Double-Blind Clinical Trial," *Journal of Chronic Diseases* 18 (1965): 367-77; R. N. Miller, "Study on the Effectiveness of Remote Healing," *Medical Hypotheses* 8:5 (May 1982): 481-90; Daniel P. Wirth and Barbara J. Mitchell, "Complementary Healing Therapy for Patients with Type I Diabetes Mellitus," *Journal of Scientific Exploration* 8:3 (1994): 367-77. Z. Bentwich and S. Kreitler, "Psychological Determinants of Recovery from Hernia Operations." Paper presented at the Dead Sea Conference, Tiberias, Israel (June 1994) and cited in Elisabeth Targ, "Evaluating Distant Healing: A Research Review," *Alternative Therapies in Health and Medicine* 3:6 (November 1997): 74-8.

30. William R. Parker and Elaine St. John, *Prayer Can Change Your Life: Experiments and Techniques in Prayer Therapy* (New York: Prentice Hall, 1957). Verna Carson and Karen Huss, "Prayer — An Effective Therapeutic and Teaching Tool," *Journal of Psychiatric Nursing and Mental Health Services* 17 (March 1979): 34-7.

31. Bruce Greyson, "Distance Healing of Patients with Major Depression," *Journal of Scientific Exploration* 10:4 (1996): 447-65.

used a general form of prayer to God for healthy patients and examined their resulting levels of self-esteem, anxiety, and depression.[32] The third study measured the recovery rates of people in an alcohol abuse and dependence program after one group received prayer from Jewish and Christian volunteers.[33] In all three studies, no significant differences were found between those prayed for and those in the control groups.

From a purely scientific perspective, no clear conclusions can be drawn from these results. Two systematic reviews of prayer research have come to this same conclusion.[34] These findings are frustrating for scientists, and reflect the great difficulty of designing an experiment to test something like prayer. Researchers cannot ensure that those in the control group receive no prayer as they can when testing a drug. Many other people could be praying for the patients, making it difficult to figure out who, overall, is being prayed for the most. For these sorts of reasons, some believe this type of research will never reveal clear-cut answers.[35]

These inconclusive results might be discouraging, especially for people who question whether God answers any prayers, or wonder if he arbitrarily picks and chooses which prayers to answer. However, there is a fundamental theological flaw with all of this prayer research. Research methods are designed to minimize the effects of human interactions on the therapy being examined. The goal is to ensure that only the therapy itself causes whatever changes are measured. But this assumes that the only choice involved in prayer is the human decision to pray. In this view, if prayer works, it will automatically lead to better results than not praying. But this is the case only if prayer is mechanistic, based on some energy or thought-power or unknown force.

Some prayer researchers do indeed view prayer as an impersonal force to be used by humans to meet their needs. The LeShan method of prayer is a meditative technique whereby "the healer accesses a state of consciousness"

32. Seán O'Laoire, "An Experimental Study of the Effects of Distant, Intercessory Prayer on Self-Esteem, Anxiety, and Depression," *Alternative Therapies in Health and Medicine* 3:6 (November 1997): 38-53.

33. Walker, "Intercessory Prayer," pp. 79-86.

34. L. Roberts, I. Ahmed, S. Hall and C. Sargent, "Intercessory Prayer for the Alleviation of Ill Health," in *The Cochrane Library* [database on disk and CD-ROM]. The Cochrane Collaboration. Oxford: Update Software, 1997, Issue 4 (updated quarterly). John A. Astin, Elaine Harkness, and Edzard Ernst, "The Efficacy of 'Distant Healing': A Systematic Review of Randomized Trials," *Annals of Internal Medicine* 132:11 (June 2000): 903-10.

35. Keith S. Thompson, "Miracles on Demand: Prayer and the Causation of Healing," *Alternative Therapies in Health & Medicine* 3:6 (November 1997): 92-6.

which then "taps a normal human ability to stimulate a patient's natural capacity for self-healing to function more quickly and efficiently than usual."[36] Another popular author views prayer as a vibrating inner connection with the divine that gives her control over esoteric powers.[37] For others, prayer is "some form of mind-to-mind communication between patient and practitioner or some form of previously undescribed energy transfer."[38] Another researcher defines prayer "as any purely mental effort undertaken by one person with the intention of improving physical or emotional well-being in another."[39] Some of the researchers equate prayer with sorcery, shamanism, psychic healing, and telepathy.[40]

Probably the most popular proponent of prayer for healing is Larry Dossey. For him, prayer is more of a general attitude of leaving things in the hands of fate, or some universal consciousness — what he calls "prayerfulness." He is deliberately vague and broad in his definition that "prayer is communication with the Absolute."[41] However, he claims that "the most common image of prayer in our culture is something like this: 'Prayer is talking aloud or to yourself, to a white, male, cosmic parent figure who prefers to be addressed in English.'"[42] He acknowledges that his view of prayer is "far different" from the "old biblically based views of prayer."[43] He claims that biblical prayer arises from a worldview which "is now antiquated and incomplete" and constitutes a "uniquely 'pathological mythology.'"[44]

Christians can welcome the renewed attention given to prayer, but we must be clear about what we mean by prayer. If we have the opportunity to get involved in prayer research, we must be very careful about what will be of-

36. Greyson, "Distance Healing," p. 449.

37. Rosemary Ellen Guiley, *Prayer Works: True Stories of Answered Prayer* (Unity Village, MO: Unity Books, 1998).

38. Sicher, "A Randomized Double-Blind Study," p. 362.

39. Targ, "Evaluating Distant Healing," 74.

40. Marilyn Schlitz and William Braud, "Distant Intentionality and Healing: Assessing the Evidence," *Alternative Therapies in Health & Medicine* 3:6 (November 1997): 62-73; Larry Dossey, *Be Careful What You Pray For . . . You Just Might Get It: What We Can Do About the Unintentional Effects of Our Thoughts, Prayers, and Wishes* (New York: HarperSanFrancisco, 1997), pp. 11-13.

41. Larry Dossey, "Prayer, Medicine, and Science: The New Dialogue," in *Scientific and Pastoral Perspectives on Intercessory Prayer: An Exchange Between Larry Dossey, M.D. and Health Care Chaplains*, ed. Larry VandeCreek (New York and London: Harrington Park Press, 1998), p. 10.

42. Dossey, "Prayer, Medicine, and Science," p. 9.

43. Larry Dossey, *Healing Words: The Power of Prayer and the Practice of Medicine* (New York: HarperSanFrancisco, 1993), pp. 6-7.

44. Dossey, *Healing Words*, p. 7.

fered. We are clearly instructed in Scripture to have nothing to do with magical practices, divination, or sorcery (Deut. 18:10-14). Many of the practices called prayer for healing fall into these categories. Yet many people will be seduced into trying these occult practices when they are lumped together under the term "prayer."

For Christians, prayer is talking to God and listening to his answers — usually provided through his Word, the Bible. Requests made to God in prayer should, according to the Bible, always be saturated with humility. We should keep in mind that we are approaching the all-knowing, all-powerful God of the Universe, talking to him about what he knows is best in a particular situation: "This is the confidence we have in approaching God: that if we ask anything *according to his will*, he hears us" (1 John 5:14, emphasis added).

Yes, we should pray for healing, as we are told in James 5:13-16.

> Is any one of you in trouble? He should pray. Is anyone happy? Let him sing songs of praise. Is any one of you sick? He should call the elders of the church to pray over him and anoint him with oil in the name of the Lord. And the prayer offered in faith will make the sick person well; the Lord will raise him up. If he has sinned, he will be forgiven. Therefore confess your sins to each other and pray for each other so that you may be healed. The prayer of a righteous man is powerful and effective.

But we should not approach prayer as either a theological right, or a medical therapy. We have no biblical guarantee that our prayers will be answered. Paul prayed for the removal of the thorn in his flesh, widely regarded as a physical ailment. He tells us:

> Three times I pleaded with the Lord to take it away from me. But he said to me, "My grace is sufficient for you, for my power is made perfect in weakness." Therefore I will boast all the more gladly about my weaknesses, so that Christ's power may rest on me. That is why, for Christ's sake, I delight in weaknesses, in insults, in hardships, in persecutions, in difficulties. For when I am weak, then I am strong. (2 Cor. 12:7-10)

Jesus Christ does not offer physical healing for every ailment. He does not provide perfect health for everyone who prays to him. Thus, the inconclusive research results are completely compatible with the biblical view of prayer. God's will is involved, and we cannot control for that in research. The results also count as evidence against the view that prayer is an impersonal energy. If it was, we would expect consistent results.

Within much of alternative medicine, prayer and spirituality are viewed as impersonal, mechanistic, and utilitarian. Larry Dossey reveals this in responding to those who see prayer research as scientifically inappropriate. "Yet, if events occur in controlled laboratory studies, as suggested by evidence cited below, these happenings presumably follow natural law and are not considered miraculous."[45] Prayer that is not miraculous? Faith that does not involve God? These do not pass the test of 1 John 4:1-3!

While we must be critical and discerning, recent interest in spirituality creates many opportunities. Christians can give up too easily, forgetting what God can do if we make ourselves available to him. We may have the opportunity to speak to physicians at a secular conference about Christian prayer, or have our theological concerns reviewed in the *Annals of Internal Medicine* — both of which have been the experience of this author.[46] Such opportunities remind us that God is at work in the world. Our responsibility is to engage our culture, making our faith clearly visible.

Christian nurses and doctors know their patients need the comfort of Jesus Christ. Students who are hurting because their marriages are falling apart or because they are too depressed to study could benefit from prayer. But the fiduciary relationship complicates matters. Those in positions of power must be careful with those who are vulnerable to coercion. Using scientific studies to validate "prescribing" prayer is going too far. It gives science unwarranted authority in religious matters.

If the patient initiates a spiritual conversation, we should respond. Alternative therapists have the reputation of caring more for their patients because they take more time to talk. Five years ago my wife and I took our two-year-old son to the emergency room with a gaping laceration to his face from a golf club. The plastic surgeon told us he would do his best, but the healing was in God's hands. My wife and I latched onto his statement, and asked if he would pray with us before starting to stitch. He left the room and said he would return after we prayed. His presence, even if in silence, would have meant so much to us.

In fact, health care professionals can do more than respond to invitations from patients; they can take the initiative in spiritual matters. Health care professionals need to be open and honest. It should be clear that a suggestion of prayer is not based on our professional role, but is a matter of per-

45. Larry Dossey, "Prayer and Medical Science: A Commentary on the Prayer Study by Harris et al. and a Response to Critics," *Archives of Internal Medicine* 160 (June 2000): 1735-8.

46. John A. Astin, Elaine Harkness, and Edzard Ernst, "The Efficacy of 'Distance Healing': A Systematic Review of Randomized Trials," *Annals of Internal Medicine* 132:11 (June 2000): 903-10.

sonal conviction. Hopefully, our patients and students will already have a sense of this, so long as our own spirituality is authentic and we emit "the aroma of Christ" (cf. 2 Cor. 2:15).

The temptation to make faith impersonal is strong. Many have turned away from the institutionalized church because the prescription of showing up on Sunday morning and having the expert dispense a dose of religion doesn't work. Ironically, many are turning to doctors and nurses as the new dispensers of faith and spirituality in times of need. Scientific studies supposedly give them the authority to do this. But once again, faith is being institutionalized and made impersonal.

Instead, God wants us to get involved personally in people's lives. He wants us to build meaningful relationships with one another so that we can help one another grow spiritually and comfort one another. The faith that is required is the belief that God exists and that he rewards those who earnestly seek him (Heb. 11:6). The reward is salvation, and the assurance of eternal life where there will be no more tears, pain, illness, or death (Rev. 21:4). Meanwhile, we are guaranteed spiritual growth, if we follow God's means of growth. Such growth will lead, not to perfect health, but to the ability to deal with whatever life brings. In many ways, this is what those who seek alternative medicine are looking for. They want what Paul found in Jesus Christ:

> I have learned to be content whatever the circumstances. I know what it is to be in need, and I know what it is to have plenty. I have learned the secret of being content in any and every situation, whether well fed or hungry, whether living in plenty or in want. I can do everything through him who gives me strength. (Phil. 4:11-13)

Preventing AIDS and STDs

MARY B. ADAM

On the playground of a local grade school one hears the noise of children running and laughing. Boys are playing soccer and girls are jumping rope. The jumping rhyme the girls recite matches the cadence of the rope hitting the pavement. Rope twirlers and jumpers all synchronized with the rope begin to chant, "First comes love, then comes marriage, then comes Suzy with the baby carriage." The next girl in line readies herself to jump and the echo comes again, "First comes love, then comes marriage, then comes Sally with the baby carriage." The refrain repeats each time with a new name until amid shouts and giggles someone stumbles and the cadence stops momentarily, only to resume again. While little girls are always making up new jump rope rhymes, some undoubtedly with the new Britney Spears lyrics, these old rhymes seem to surface again and again, calling to mind an order that for many has become passé: first love, then marriage, then sex. However, the rhyme has a new order on today's college campuses. First sex, then maybe love, then maybe live together.

The sexual revolution that took place on college campuses of the 60s planted seeds that grew into the sex-obsessed culture of today. In the new millennium multiple partners is seen as the norm. In the new campus order sex frequently isn't even referred to as "making love" — it's called "shagging," "bonking and screwing," or "hooking-up." These descriptions of sex on campus are written by recent graduates and found in books like *Sex on Campus: The Naked Truth about the Real Sex Lives of College Students*, 1997, in which the author states:

> In recent decades, the students at small colleges seem to have moved away
> from the whole concept of dating. . . . The favored approach is just to play it

131

cool and wait until you see the person again to develop your relationship further. . . . Hooking up: you were almost certainly acting on physical attraction, not a well-formed physical attachment, and there was no risk for either of you. You're under no obligation to date each other or call each other; nor should you expect to be called or dated. . . . Ball-and-chain rating: you should never get so drunk that you do something you didn't want to. In reality, however, a great many college hookups occur when both parties are sloshed . . . [but] if you realize almost immediately after you finish having sex that this will definitely be a one-time event and you really don't want to pursue any relationship — even a purely physical one — with this person, try not to sleep through the night with the person. It may seem awfully awkward, and it may be late at night, but get dressed, say "thank you for a wonderful evening," and go home. Leaving someone with whom you've just traded bodily fluids can seem strange, rude, and inconsiderate, but at least you'll have the knowledge that you were being honest, and it will make things less complicated down the road.[1]

Wendy Shalit, a 24-year-old graduate of Williams College, describes hookup this way:

Hookup is my generation's word for having sex (or oral sex) or sometimes for what used to be called "making out." The hookup connotes the most casual of connections. Any emotional attachment deserves scorn and what *Sex on Campus* calls a dangerously high "ball and chain rating" . . . it's such a strange expression, hooking up, like airplanes refueling in flight. Not just unerotic but almost unintimate.[2]

Another author uses even stronger words:

Most of my contemporaries no longer make love. They shag, bonk and screw — quickly and anonymously, lovelessly. The generation searching for intimacy more pitifully than any other in history has taken the central sacrament of interpersonal intimacy and killed it dead. We have the dubious privilege of living in the culture that is presiding over the death of eroticism.[3]

Casual sex and the concomitant multiplicity of sexual partners reflect a huge shift in sexual attitudes. This shift has been documented by the Centers

1. L. Elliot and C. Brantley, *Sex on Campus: The Naked Truth about the Real Sex Lives of College Students* (New York: Random House, 1997), 49-54.
2. W. Shalit, *A Return to Modesty* (New York: The Free Press, 1999), 28.
3. M. Starkey, *God, Sex, and Search for Lost Wonder* (Downers Grove, IL: InterVarsity Press, 1997).

for Disease Control's (CDC) Youth Risk Behavior Surveillance, which shows that the age of first intercourse continues to drop. Over eight percent of high school students report first intercourse before age 13. More than 60% of high school seniors report having been sexually active, and of those, over 20% report four or more sexual partners. In addition, 24% of high school students report using alcohol or drugs during sexual intercourse.[4]

Unfortunately, this shift in sexual attitudes and practice has produced an epidemic of sexually transmitted diseases (STDs) that is unprecedented.[5] STDs are major public health concerns. Over thirty percent of women on college campuses are infected with human papillomavirus (HPV), a sexually transmitted infection that is the precursor of cervical cancer.[6] Chlamydia, an infection that can cause painful urination, pelvic inflammatory disease and infertility, can be found in about 1 in 10 high school students; and the CDC estimates that 1 in 250 people in the United States have HIV infection.[7] Public health messages intended to stem the tide of STDs have focused on encouraging individuals to use condoms. This emphasis on self-protection has been echoed in condom advertisements shown in recent periodicals. A full-page ad for Lifestyles condoms in the April 2000 issue of *Family Planning Perspectives* magazine shows a picture of a pensive-looking young woman with the following caption:

"Why Every Woman Should Use a Condom"

We don't need a condom because he says he loves me. Life would be so uncomplicated if love did conquer all and trust was the only thing a woman needed for protection. But the truth is those women are totally at the mercy of a man's honesty and sincerity when it comes to safer sex.

Studies reveal that less than 20% of all sexually active men use condoms 75% of the time. And that 50% of the men questioned would lie about their sexual past. To compound the problem, many ethnic and minority cultures

4. L. Kann, et al., *Youth Risk Behavior Surveillance — United States, 1999* (Centers for Disease Control and Prevention, 2000).

5. J. McIlhaney, "Sex in America: Past, Present, and Future," in *The Reproduction Revolution: A Christian Appraisal of Sexuality, Reproductive Techniques, and the Family,* ed. J. Kilner, P. Cunningham, and D. Hager (Grand Rapids: Eerdmans, 2000).

6. R. D. Burk, et al., "Sexual Behavior and Partner Characteristics Are the Predominant Risk Factors for Genital Human Papillomavirus Infection in Young Women," *Journal of Infectious Diseases* 174:4 (1996): 679-689.

7. D. A. Cohen, et al., "A School-based Chlamydia Control Program Using DNA Amplification Technology," *Pediatrics* 101:1 (1998): E1.

consider the use of condoms to be forbidden or unmanly. This behavior and attitude has put women at great risk.

An unwanted pregnancy can ravage a young woman's life emotionally, economically, and educationally. 56 million Americans are infected with a sexually transmitted disease and these diseases have the greatest effect on women. In 1992, the ratio of women to men infected with AIDS was 1:7; by the year 2000, it will be 1 to 1. Heterosexual women have become the fastest growing AIDS segment of the population.

It should now be obvious that "trust me" and "I love you" don't quite add up to a safe sex guarantee. If a woman wants to be totally secure, she should practice abstinence. But . . . if she chooses to be sexually active, she must have an alternative. Latex condoms, when used properly and consistently, provide effective barrier protection for the prevention of AIDS, STDs, and unwanted pregnancy.

From now on, when it comes to safer sex, every woman must be convinced that "I love you" should mean, "I'll use a condom."

Once again, the old jump rope rhyme and the new reality don't mesh. Love doesn't mean marriage; it means, "I'll use a condom."

Relationships at Risk

This collision of old and new paradigms results in confusion about the meaning of sex and the meaning of relationships. However, many couples long for closer, more intimate relationships, not just casual sex. Yet things really get confused when one adds HIV prevention into this mix. The evidence indicates that individuals in close personal relationships do not use any effective HIV prevention practices even though the risk of contracting HIV seems to be rising.[8] Why would such seemingly foolish behavior be so common? Recent research gives some insightful answers.

Frequently, as relationships mature, couples stop using condoms and start to use oral contraceptives. This switch away from condoms appears to be symbolic and meaningful to the individuals in relationships. For some, ceasing to use condoms appears to indicate an end to worry about cheating, prior partners, and poor communication about sexual protection. It symbolizes the beginning of a trusting long-term relationship. One participant has stated

8. S. J. Misovich and W. A. Fisher, "Close Relationships and Elevated HIV Risk Behavior: Evidence and Possible Underlying Psychological Processes," *Review of General Psychology* 1:1 (1997): 72-107.

condom use in a long-term relationship "almost proves . . . a lack of trust." Couples commonly feel glad not to have to use condoms any more. Indeed, many people just don't like using condoms, especially men. Unfortunately, this change to birth control pills plays a major role in couples ceasing to practice HIV prevention.[9]

One reason for switching from condoms to pills without mutual HIV testing to determine if a partner is safe is a belief that intimate relationships are inherently safe even though the objective factors used by couples to judge safety (e.g., trusting, liking, or knowing the partner) are irrelevant to the partners' actual level of HIV risk. There is a common belief that "known partners are safe partners." The more a relationship progresses, the more difficult it is to think that the partner whom one "loves" ever engaged in behaviors that could result in HIV infection, could presently have a life-threatening illness, or could transmit a deadly virus. The old adage "love is blind" seems still to be true. Another incorrect assumption that facilitates risky behavior in relationships is that trusted partners are safe partners. This means that trust, which develops over time, is seen as a substitute for condoms or HIV testing. This irrational substitution occurs not only among college students but in other populations as well. Simply put, when relationships are going well, few partners consider the possibility of HIV risk within the relationship.[10]

The awkwardness many couples feel in conversing about HIV and HIV prevention presents another major difficulty in realistically assessing a partner's level of HIV risk. Often as a relationship deepens, communication improves. However, communication in relationships typically occurs in such a way as to enhance trust, not undermine it. Conversations which decrease certainty or intimacy are usually avoided. Addressing issues of past sexual partners, previous STDs, and risk for HIV are the types of conversations that can move toward uncertainty. A study by Hammer et al., entitled "When Two Heads Aren't Better than One: AIDS Risk Behavior in College-age Couples," found that one third of female and one half of male participants were concerned that bringing up the subject of HIV testing was the equivalent of accusing their partner of cheating or sleeping around. One woman subject

9. J. C. Hammer, et al., "When Two Heads Aren't Better than One: AIDS Risk Behavior in College-age Couples," *Journal of Applied Social Psychology* 26:5 (1996); M. L. Cooper, et al., "Motivations for Condom Use: Do Pregnancy Prevention Goals Undermine Disease Prevention among Heterosexual Young Adults?" *Health Psychology* 18:5 (1999); C. J. Pilkington, "Is Safer Sex Necessary with a 'Safe' Partner? Condom Use and Romantic Feelings," *Journal of Sex Research* 31:3 (1994).

10. Misovich and Fisher, "Close Relationships."

summarized the difficulty with discussing HIV by saying, "If you're not comfortable talking about basic feelings, you're not going to say you want an AIDS test." Another stated, "We never ever talk about it" and "I don't think I could ask him." Discussions about condom use or HIV testing under these conditions are often seen as a betrayal of trust. Even when partners do become intimate enough to discuss issues of HIV prevention, they frequently feel constrained to accept whatever their partner says in order to prove their level of trust.

Many couples consider HIV testing as a useful way to protect against HIV. Unfortunately, they commonly reject actual testing. In the study of college-age couples just cited, seventy-five percent had not asked their partners to be tested for HIV even though they had often considered it. Trust was the issue. Some participants expressed concern that bringing up the subject of HIV testing might make their partner feel suspected or mistrusted, or that they would regard the request as an insult. Others worried that HIV testing might lead to an imbalance in the relationship, with one partner blaming the other for the necessity of getting tested because "you've had so many partners." Some couples expressed concern that HIV testing would be viewed as a sign that the relationship was becoming "too serious." Finally, participants tended to avoid testing because they believed that if either partner tested positive, it would cause the relationship to end. The researchers quoted one person as saying, "It's scary. You don't want to think of your boyfriend having AIDS. You're out a boyfriend."

The findings of current research seem to indicate that mutual trust, intimacy, and sharing within relationships are issues so critical for couples that a danger as large as HIV can be ignored when it threatens these aspects of a relationship. This seems to be the case for both heterosexual and gay male couples. What many couples are actually showing is that they are strongly disposed to reject safer sex practices which are perceived to inhibit intimacy or violate perceptions of trust.[11]

This outlook poses a tremendous dilemma for prevention policy and public health. The Centers for Disease Control estimate that currently 1 in 250 Americans is infected with HIV, and the numbers are growing. Policy regarding HIV prevention has been built on a foundation of condom use for

11. Hammer, "When Two Heads Aren't Better"; Misovich and Fisher, "Close Relationships"; S. J. Misovich and W. A. Fisher, "The Perceived AIDS-preventive Utility of Knowing One's Partner Well," *Canadian Journal of Human Sexuality* 5:2 (1983): 83-90; J. D. Fisher, W. A. Fisher, and T. E. Malloy, "Changing AIDS Risk Behavior: Effects of an Intervention Emphasizing AIDS Risk Reduction Information, Motivation, and Behavioral Skills in a College Student Population," *Health Psychology* 15:2 (1997).

self-protection. Obviously, the prevention message, "use a condom to protect yourself," is failing to work. What is the answer? Is it possible to make condoms non-threatening to the relationship? Almost certainly it is not, for all the reasons stated. Is it possible to alter couples' perceptions of HIV testing? Currently, HIV testing is free and available through any county health department. It remains underutilized at best. Unfortunately, removing cost and accessibility barriers does not address the threat HIV testing may bring to the relationship. So is there an answer?

Perhaps there is, but one must look in a radically different direction. Improved HIV prevention for these couples must be directed at the couple, not the individual, and self-protection cannot be the primary thrust of the message. Hammer et al. summarize the difficulties of applying risk reduction methods for couples in intimate relationships: "those in intimate relationships believe it is highly unlikely that their partner could be HIV positive, but highly likely that initiating condom use or HIV testing could damage their relationship, which is hardly a favorable cost-benefit ratio. . . . Designing interventions to foster HIV prevention in couples should be a priority."

Designing an HIV prevention message that is effective for couples who are seeking greater intimacy must be grounded in the idea of care and concern for the other more than for oneself. As one participant in the Hammer study said, "If you care about each other, it's [getting HIV tested] not that big of an issue." Caring for others more than oneself is a virtue not commonly seen in today's postmodern milieu where underlying individualism profoundly colors the way people view all of life and especially sexual relationships. Caring for others more than oneself requires a redefined understanding of what relationships are all about. It is a notion rooted in a Christian worldview. The Christian worldview has not carried much weight for many persons working in HIV prevention, but a paradigm centered in something besides individual autonomy is required to address the HIV prevention issues among the couples who are seeking intimacy and are rejecting current safer sex methodologies. One asks the question, can different types of relationships be taught to a generation that seems flooded with a "me first" orientation?

Improved Prevention Strategies

Curricula that address relationship issues do exist and are being evaluated for use with high school students. Relationship training in high school makes perfect sense if one wants to alter HIV risk behaviors, since high school is the time when youth are exploring relationships and many are becoming sexually

active. Curricula like *WAIT Training*, put out by Friends First, emphasize relationships.[12] One section in this curriculum discusses stages of a healthy relationship and identifies a variety of sexual foreplay and intercourse behaviors
as appropriate only for marriage. This curriculum also provides information
on how to identify if a couple is compatible and addresses issues of commitment in terms youth can understand.

Other strategies that emphasize character education have also been developed and are currently employed. One promising curriculum entitled *Love
and Life at the Movies* emphasizes education for character through the film
classics.[13] It begins with *Our Power to Love,* a four-film series that focuses on
introductory lessons of character and human relationships. This curriculum
is being widely implemented in Virginia and is being evaluated using federal
Title V funding. There are also prototypes of community intervention programs designed to prevent adolescent pregnancy, like M. L. Vincent's program in South Carolina. In such programs, teachers, ministers, parents, and
community leaders are trained in a cohesive manner about relationships and
sexual topics. In Vincent's program, the primary behavioral objective is to
postpone initial voluntary sexual intercourse among never-married teens and
preteens. The results show a dramatic decrease in unwed teen pregnancy.[14] It
is possible that community-based programming directed at establishing solid
virtuous relationships and postponing initial voluntary sexual intercourse
would have similar impact in reducing HIV risk behavior. Those public
health officials in charge of HIV prevention would do well to look closely at
some of these programs that emphasize relationships as well as postponing
age of first intercourse.

In addition, the role of the church in HIV prevention and sexuality education has received little attention. The initial reluctance of those in the
church to confront the issue of HIV has changed. Indeed, there are notewor-
thy examples of church and para-church groups reaching out to youth with a
message about strong relationships and avoidance of STDs. Two cases in
point are the Prevention Services arm of the Tucson Crisis Pregnancy Center
and Carondelet Health Network's Youth Sexuality Program. It is individuals
in the church that can uniquely model caring relationships based on having a
redeemed relationship with God. Yet the church could be positioned to take a
broader role.

---

12. *WAIT Training* (Longmont, CO: Friends First, 1996).
13. O. McGraw, *Love and Life at the Movies* (Educational Guidance Institute, 2000).
14. M. L. Vincent, A. F. Clearie, and M. D. Schluchter, "Reducing Adolescent Pregnancy through School and Community-based Education," *JAMA* 257:24 (1987): 3382-86.

A recently published study entitled "The African-American Church: A Potential Forum for Adolescent Comprehensive Sexuality Education" identifies several strengths religious organizations have in educating young people about sexuality.[15] There is no other institution that reaches a larger portion of the population of Americans than faith-based organizations. Faith-based organizations have a long history of addressing health issues, a priority that stems from Christ's care and concern for the sick. Churches, especially African-American churches, have a long tradition of being not only spiritual growth and development centers but also centers of political and civic activity. The church has also long been recognized as having a critical role to play in moral guidance. As private institutions, churches have important advantages over public schools in dealing with controversial topics like sex education. For all these reasons, the church is uniquely positioned to address HIV prevention and the aspects of relationships that would promote truly safe sex.

"The African-American Church" article also points out that the highest priority issues of the surveyed clergy leaders are drugs, violence, HIV/AIDS, pregnancy, and alcohol. Seventy-six percent have discussed one or more of these issues in church. All of the respondents have requested additional health seminars for adolescents, though some clergy (30%) exclude some topics (i.e. anal sex, bisexuality, homosexuality, masturbation, and oral sex). This sample of African-American clergy is looking for an increased role in teaching their young people about prevention, including HIV prevention. The church definitely has a message about sex and relationships that needs to be heard. Mike Starkey said it well in his book, *God, Sex and the Search for the Lost Wonder* (1997):

> We live in a culture that actively discourages personal integrity and maturity in relationships. . . . Many books about sexuality that are aimed at Christian teens give lists of do's and don't's. This seems to miss the point and is in many ways counterproductive. For one thing, human nature is such that on seeing a list of prohibited pleasures, our immediate desire is to try out every single one of them! We ask what is so special about this thing that they are trying to keep us from doing. Second it [dictating behavior] falls into the same trap as our sex-sated culture. It discourages maturity in relationships and personal responsibility. It lays down laws, telling what we ought to be doing, rather than inviting us to reflect on what will help us grow as integrated, fulfilled people and to discover the role one's sexual and

15. T. Coyne-Beasley and V. Schoenback, "The African-American Church: A Potential Forum for Adolescent Comprehensive Sexuality Education," *Journal of Adolescent Health* 26:4 (2000): 289-93.

other relationships play in the process of identity shaping. So the message to those outside a lifetime covenant relationship is an invitation to intimacy and relationship, no less so than to those who are married. But the ways this is expressed will be different. It will involve rediscovering the lost art of vulnerable, self-giving friendship, with people of the opposite sex and the same sex, and rediscovering the touch of affection. Most of all it will involve learning to be a whole, integrated person, resisting a society that severs body from mind, sexuality from commitment. Our bottom line must be a challenge to the strange credo of our sexual culture — that even bad copulation is better than good friendship.

Does the church have a role in teaching HIV prevention to couples who are already seeking something better than "bad copulation" but are unwilling to accept a biblical definition of morality? Even as the church seeks to model anew an old standard, there are many who reject sexual fidelity and the lifelong commitment of marriage. Many in the current generation would respond to those who promote a biblical definition of sex by saying, "What right do you have to tell me how to run my sex life? I'm happy with it." At this juncture, the church and public health officials share a common rejection of their respective messages. The public health message about condom use is rejected and the church's message about relationship fidelity and marriage is rejected. Yet at the same time, many couples are rejecting what they call casual sex, and casual sex is where they are more likely to use a condom to protect themselves. They've "been there, done that" and find casual sex lacking something. As Starkey puts it: "A whole generation of young adults is doing what it likes, but they no longer like it very much. Xers have discovered the emptiness of casual sex, the boredom of commitment-free intimacy, the agonizing loneliness of a life lived without openness and vulnerability of true friendship. They are longing for a fresh vision."

The public health community and the church can agree that a fresh vision is needed. Current methods aren't working. STDs, HIV, and sexual risk behavior are rising in parallel. Both institutions need to continue to identify innovative ways to reach couples whose behavior puts them at risk for a deadly disease. These institutions approach this dilemma from different directions: public health points to risk reduction and the church to risk avoidance. Nevertheless, both are needed in a fallen world. There is plenty of room for working together on a problem with the proportions and ramifications of HIV. Realistically, the church may not have much impact on those couples who currently reject biblical relational standards. Yet the church is in a much better position than public health to deal with those wounded in the sexual jungle. First, those in the church believe there is a possibility for radical change in a person's life. The God who created can also restore, overcoming the effects of sin and the Fall. Second, God's people are called to carry out a ministry of hope and healing through acts of charity and service. The local health department has no response to the person who is struggling with the pain of infertility caused by an STD. The local health department has medicine to give the individual who got an STD despite using a condom if the infection is treatable, but nothing to offer for those infections with no cure. Here the church has much more to offer than the health department. Medicine, no matter how effective in curing a given disease, doesn't always cure the greatest hurts. The church is the place where there is true expertise in rebuilding broken people and broken relationships.[16]

16. Special thanks to Anne Burnson for her editorial assistance.

The Ethic and Spirit of Care

DIANN B. UUSTAL

And when he saw him, he had compassion. So he went to him and
bandaged his wounds . . . and took care of him.

Luke 10:33, 34[1]

We all know what it means to care for or about someone or something. But
just what is *care?* What does it mean to be *caring?* What is the *ethic of care?*
What distinguishes professional, therapeutic caring from the care we give one
another as family members and friends?

Do care and caring *influence healing?* What do people *do* when they care
for one another? How do you know you've been *cared for* or you've given *good
care?*

This chapter offers the opportunity to explore these and other related
questions.

Care & Caring

Most of us are familiar with the terms *care* and *caring* and maybe even *therapeutic care* and the *ethic of care.* However, briefly defining these words and
concepts at the beginning of this article will clarify their meaning and use.

The concepts of care and caring have several attributes that have been

---

1. *Women of Destiny Bible* (2000), The Holy Bible, New King James Version (Nash-
ville: Thomas Nelson, 2000).

discussed in the literature by numerous philosophers, anthropologists, caregivers, and researchers. The first attribute is described in the work of Heidegger,[2] who suggests that "caring is a natural state of human existence, a way that we relate to others and our world, and a way of being in the world." This description of care is supported in the research of Nel Noddings, who also believes that caring is a "natural sentiment." She states that caring is not necessarily regarded as moral behavior, but rather is a natural disposition of being human. She also says that "to care may mean to be charged with the protection, welfare, or maintenance of something or someone."[3] Willard Gaylin also defined caring as "behavior that is naturally programmed in human nature that can be influenced positively or negatively by one's environment."[4]

A second attribute of caring is its status as a precondition for caring behaviors. Anne Griffin[5] agrees with the idea that caring exists as an integral feature of human growth and development before one can care about things, others, or oneself and before the behaviors associated with caring take place.

A third attribute of caring is that it is a moral phenomenon. Caring occurs in cultures so that human needs are met. Caring professions exist because they can offer a structured way to meet basic needs. Since caring serves the needs of others and is associated with the ideals and values of communities, it is a moral phenomenon. Griffin has analyzed caring and divides the meanings of caring into two main groups. The first group is what she calls the "taking-care-of activities" associated with caring and includes terms such as consideration, concern, guidance, protection, and serving needs.[6] The second group of meanings centers around the feelings and attitudes that are associated with the activities of caring. Griffin considers these two groups of meanings to be nonmoral aspects of caring and believes the moral aspects of caring involve the motivation and character of the caregiver. Milton Mayeroff also describes caring as a moral endeavor based on the motivation to protect and maintain a person's welfare. He distinguishes caring from feeling or emotions and describes caring as "a process, a way of relating to someone that involves development in time through mutual trust and a deepening and qualitative

2. Martin Heidegger, *Being and Time,* trans. J. Macquarrie and E. Robinson (New York: Harper & Row, 1962).

3. Nel Noddings, *Caring: A Feminine Approach to Ethics and Moral Education* (Berkeley: University of California Press, 1984), p. 9.

4. Willard Gaylin, *Caring* (New York: Avon Books, 1976).

5. Anne Griffin, "A Philosophical Analysis of Caring in Nursing," *Journal of Advanced Nursing* 8 (1983): 289-295.

6. Griffin, "Philosophical Analysis," p. 290.

transformation of the relationship. . . . To care for another person in the most significant sense, is to help him [sic] grow and actualize himself."[7]

Madeleine Leininger, a nurse and cultural anthropologist, maintains that care is an essential human need for the full development, health maintenance, and survival of human beings in all world cultures.[8] She describes care as essential for the survival of the human race because it is used in all cultures to protect human beings and to reduce stress and conflict. Leininger defines caring as "those human acts and processes that provide assistance to another individual or group based on an interest in or concern for that human being(s), or . . . meet an expressed, obvious, or anticipated need."[9] She states that caring refers to "the direct (or indirect) nurturant and skillful activities, processes, and decisions related to assisting people . . ." and that it includes ". . . behavioral attributes which are empathetic, supportive, compassionate, protective, succorant, and educational. . . ." Caring focuses on "the needs, problems, values, and goals of the individual or group being assisted."[10]

Jean Watson, also a nurse, views caring as a value and an attitude that becomes an intention or commitment and becomes apparent in specific behavior. She states that "caring calls for a philosophy of moral commitment toward protecting human dignity and preserving humanity," and believes that caring "manifests itself in concrete acts."[11]

Physician-ethicist Edmund Pellegrino describes caring as compassion or being concerned for the other person, doing for others what they can't accomplish on their own, directly caring for the medical problem, and competently taking care of the individual from both a personal and a technological perspective. He does not separate these four aspects of caring and sees caring as a moral obligation of all health care professionals.[12]

There are many multidisciplinary health care professionals involved in care and caring for individuals: physicians, nurses, social workers, pastoral caregivers, physician assistants, physical therapists, sonographers, and administrators, to name just a few. Each discipline focuses on different goals

7. Milton Mayeroff, *On Caring* (New York: Harper & Row, 1971), p. 1.

8. Madeleine Leininger, *Transcultural Nursing: Concepts, Theories & Practices* (New York: Wiley & Sons, 1978).

9. Leininger, *Transcultural Nursing*, p. 46.

10. Leininger, *Transcultural Nursing*, p. 4.

11. Jean Watson, *Nursing: Human Science and Human Care: A Theory of Nursing* (Norwalk, CT: Appleton-Century-Crofts, 1985), pp. 31, 32.

12. Edmund Pellegrino, "The Caring Ethic: The Relation of Physician to Patient." In *Caring, Curing, Coping: Nurse, Physician and Patient Relationships*, ed. Ann Bishop and Judy Scudder (Tuscaloosa, AL: University of Alabama Press, 1985), pp. 10-12.

and objectives for patient care and approaches a patient/client from a slightly different perspective. While all the health care professions offer care and value caring, Leininger states that "care is the essence and the central, unifying, and dominant domain to characterize nursing," and that "care is the unique and major feature that distinguishes nursing from other disciplines." She further accentuates that it is care that is the "skill and ethic that characterizes nursing."[13]

Therapeutic care refers to a learned art and science that focuses on and incorporates individualized actions, processes, and skills intended for the benefit of a person or a group. Its goals include recovery and restoration from illness, the maintenance, improvement, or promotion of health, rehabilitation, pain management and palliation, and the preparation for a meaningful death. Therapeutic care is based on a whole person or holistic approach that includes physical, emotional, spiritual, relational/social, and intellectual aspects of caring for the individual. It involves being committed to expressing attitudes and taking action on behalf of individuals that supports their healing and well-being, decreases their pain, and meets their obvious or anticipated needs during difficult circumstances.[14] Therapeutic caring requires personal participation and committed involvement as well as advocacy and effective, ethical action from health care professionals.

In general, people intuitively know and appreciate that care and caring are essential to effective relationships, good health, and recovery. To further articulate the importance and essence of care and caring, several assumptions found throughout the literature can be identified. Briefly, the most salient are:

- Human caring is a universal phenomenon.
- Caring is essential for human birth, development, growth, survival, and for a peaceful death.
- The expression of self-care and care for others varies in different cultures and in different care systems and in relation to specific values.
- Individuals from different cultures can identify caring and non-caring behaviors, attitudes, and practices.
- Care involves several dimensions: physical, psychological, spiritual, relational/social, intellectual, and environmental. All these dimensions must be addressed if care is to be holistic, or focused on the whole person.

13. Leininger, *Transcultural Nursing*, pp. 3, 4.
14. Leininger, *Transcultural Nursing*, p. 46.

- Care and caring are culturally derived, which requires that the caregiver be knowledgeable about the culture in order to be effective in the relationship.
- Curing cannot take place without caring, but caring is possible even if curing is not.

One final hypothesis that is alarming and evident in many cultures needs further research:

- In cultures where there is a greater dependency upon technology to give treatment, there are greater signs of depersonalized care to individuals.[15]

After reflecting on these definitions of and assumptions about care and caring, it is not difficult to appreciate that care is essential to human health and well-being in all cultures and that caring enhances our development and personal interactions. Care is a feeling of concern, regard, and respect one person has for another that is strengthened over time. Caring is a personal and professional way of relating that transforms and enhances relationships. Professional, therapeutic caring involves expressing attitudes and using learned skills to take action from a holistic perspective on behalf of individuals. These skills and actions are undertaken to support a person's well-being and healing, to decrease pain, and to meet his/her needs during difficult circumstances. Professional caring requires openness and receptivity, personal participation and committed involvement, and relatedness with people. It demands responsiveness, advocacy, and effective, ethical action from health care professionals. As Leininger reminds us, care appears to be "the hidden quality of human services that makes people satisfied or unsatisfied with health services."[16]

The Ethic of Care

To care is to engage in an ethical enterprise. Caring behaviors can and do have moral content. When we talk about the "ethic of care" what do we mean and what values and concepts are involved?

One of the key features that distinguishes the ethic of care is that the nature of the healing relationship is central. The ethic of care draws our atten-

15. Leininger, *Transcultural Nursing,* pp. 5-6.
16. Leininger, *Transcultural Nursing,* p. 8.

tion to the fact that what is happening between the caregiver and the person being cared for is vital. This relationship is important to both the patient and the caregiver, should be highly regarded, and influences healing.[17]

The ethic of care focuses on several values and concepts that influence the quality of therapeutic, healing relationships. These include compassion, concern for others, the strengthening of relationships, enhancing a sense of connectedness, improving communication, being attentive to the uniqueness of the individual and the context of the situation, taking into account the importance of feelings and how they influence decisions, and being responsive and responsible for one's actions.

In the bioethics library at Georgetown University is the sculptured face of one of the founders of the field of bioethics, Andre H. Hellegers, M.D., with the inscription of one of his famous inspiring quotes: "Compassion with competence: that is the need, and where either is missing, little good will be done." It captures what is so important about the ethic of care. It is the care and caring expressed with passion and the competency in caring of the caregiver that constitute the foundation of therapeutic, healing relationships.

A conscious honoring and demonstrated respect for individuals is highly valued in the ethic of care. Respect for a person arises out of the notion that, because we are made in the image of God, each person has inherent worth and dignity. This worth does not arise from any particular status. It is inherent in being God's creation — humans.

Respect for a person is demonstrated by paying attention to the uniqueness of each individual, his or her life "story," honoring the individual by including him/her in the decision-making process, honoring the individual's values and preferences, and acknowledging and protecting autonomy (rightly understood). Respect for an individual means the decisions that are made are centered on the patient's values and not the values and preferences of other individuals who may have more power to influence those decisions. Howard Brody captures this concept when he states, "The patient ideally has a right to a relationship that assures that s/he will be treated with respect and that medical knowledge will be used to further his/her own life plans and values."[18]

Honoring an individual's cultural identity and religious integrity is also a part of how we can show respect to a person. Culture and faith significantly influence a person's views about health, sickness, pain, vulnerability, aging,

17. Edmund Pellegrino and David Thomasma, *For the Patient's Good* (New York: Oxford University Press, 1988).

18. Howard Brody, *The Healer's Power* (New Haven: Yale University Press, 1992), p. 37.

dying and death, and after-life beliefs, and should be explored with sensitivity and attentiveness. Being nonjudgmental as a caregiver is another expression of respecting individuals.

From the ethic of care perspective, respect for persons is not limited to the professional/patient relationship. It also includes how health care professionals ought to be regarded and treated by one another. *All* health care providers — regardless of title, degree, or status in the system — are included. All should be treated with the same regard and honor and they should have a voice in workplace governance and in shaping policy.

Another ethically relevant concept in the ethic of care is shared decision making. One of the ways we show respect for an individual is that we share information and invite the person to reflect on his/her values and indicate his/her preferences. This allows the person to give informed consent or informed refusal. The best interests of the individual can be determined by being attentive to the person's "story" and values. These values can be assessed by determining the patient's value history and through the use of advance directives and help from a surrogate decision maker who speaks from the patient's perspective. Whenever it is possible, the person whom the decision impacts the most should be the primary decision maker.

Power is ethically relevant because it is unequal between patients and caregivers. Power can be unequal in terms of knowledge, freedom, and influence. Because of this known possibility of inequality, it is important to recognize that sharing the decision making and enhancing a person's ability to choose are ways we express caring for people.

Our attention is called by the ethic of care to the fact that therapeutic caring relationships ought to benefit more than burden the person. This means we must consider the balance of the burdens and benefits involved in care as well as the differences that may exist between the medical effect that is achievable and the benefits perceived by the patient. Pellegrino observes that an appropriate action is one that advances the biomedical good of the patient, the patient's concept of his/her own good, and "the good most proper to being a human."[19] Pellegrino believes that the patient's good should take precedence over other notions of good. Too often the decisions made are based on the caregiver's preferences, the family's wishes, or the system's needs, and these are not necessarily what is "good" according to the patient's values and beliefs.

The ethic of care embraces familiar ethical principles. Additional ethical obligations directly and indirectly support the goals of benefiting the pa-

19. Pellegrino, "Caring Ethic," p. 22.

tient and encouraging holistic healing. These ethical principles are commonly invoked and constitute promises we make as caregivers that are nonnegotiable in therapeutic relationships.

Sanctity of life has a broad range of meanings and attempts to convey that life has importance, worth, significance, and value and that we should have reverence for an individual's life. It does not mean life at all costs, or that biological life is an end in itself, but rather it includes the recognition that it is a means to an end or other goods as well. Medicine's primary goals of preserving and protecting life and promoting health (as well as the goals of restoration, rehabilitation, alleviating a person's pain and palliation) arise from this primary ethical obligation. *Beneficence* is the duty to do good and to benefit an individual. This obligation is the cornerstone of nursing's code of ethics and is a cornerstone value for all caregivers. *Nonmaleficence* is the obligation to do no harm. In addition it also entails preventing harm if possible, or removing it if possible when it does occur. Nonmaleficence and beneficence together are the foundation of medicine. They are values shared by all health care providers. *Veracity,* which is defined as the obligation to tell the truth, is intimately associated with the concept of trust, which is seen as essential in therapeutic, caring relationships. *Justice* refers to the equitable distribution of risks, goods, and benefits and much is written from a care-based perspective about fairness in relation to access to quality health care service and our obligation to advocate for change. Also important are *fidelity,* or the duty to keep one's promises, and *confidentiality,* or the obligation to hold as private any information entrusted in the context of special relationships such as that of the caregiver and patient. *Autonomy,* or self-rule, which is also known as respect for persons, is another ethical principle embraced by the ethic of care. It must be rightly understood, though, in a way that does not divorce the individual from family and community. While this is not a complete list of ethical principles, collectively and individually they are inherently important in the ethic of care.[20]

Another essential consideration in the ethic of care is that the patient/client should be cared for from a whole person perspective. This means ministering to the person as a whole person by considering and including the physical, emotional, spiritual, social (relational), and intellectual needs of the individual. This is easy enough to say, but even casual observation indicates that this is not characteristic of the care given in many contemporary clinical settings. Too often the caregivers respond to the person on the basis of the diagnosis, disease, or the bodily system that needs attention. In addition the multiple

20. Tom Beauchamp and Jim Childress, *Principles of Biomedical Ethics,* 4th ed. (New York: Oxford University Press, 1994).

caregivers, each with their own specialty and area of concern, tend to focus on only one part of the person. As a result, the person is not seen as an integrated, whole person or from a holistic perspective, and as a result the care the person receives is fragmented. Socrates long ago observed, "You ought not attempt to cure eyes, without head, or head without body; so you should not treat body without soul."[21] This seems to be ageless wisdom and yet a lesson that many health care systems and professionals have not mastered.

That the relationship between the caregiver and patient is predicated upon a covenant and not merely a contract is another feature central to the ethic of care. The word covenant invokes a different image, calls upon a different set of centrally motivating values that influence the relationship, and shapes our responses and behavior. Covenant involves beneficence-in-trust and is grounded in the relationship between caregiver and patient.[22] A contractual model is based on autonomy and preestablished guidelines. When caregivers are clear on what it means to be involved in a covenant with someone, we think and behave differently than we would if we were operating on the basis of a contract.

The present focus on the business of health care has even changed the use of language: *customer* has now replaced the word *patient*. The word *patient* from Latin means, "one who suffers." A person who is sick is no customer. She is vulnerable, fearful, in pain, and may lack necessary information and even energy to participate in crucial decisions. A person in this state rarely wants to be seen as or treated like a customer. The first thing that comes to mind when a person is sick, injured, or otherwise vulnerable is that she wants to be cared for. This is not the image of a customer. In addition, there is an overemphasis on health care as a purchasable *commodity*. These and other more subtle factors subsequently devalue the beneficence-in-trust and covenant nature of the caring relationship. In addition, the word *covenant* somehow reminds people of religion and church, and too often they place a limited or negative connotation on the word. Covenant should actually conjure up words like *promises, special,* and *sacred.* It is preferable to have these words describing the nature of a healing relationship than words that are appropriate in business, but that devalue the special nature of the relationship between caregiver and patient.

The ethic of care also emphasizes that the character and integrity of the caregiver are relevant. Character is the internalized inclination to behave in a certain way, and integrity can be defined as predictability. The view that char-

21. William Guthrie, *Socrates* (Cambridge: Cambridge University Press, 1971).
22. Pellegrino and Thomasma, *For the Patient's Good.*

acter counts and that integrity is relevant is of importance because the one being cared for should be able to count on certain things happening in the context of a caring relationship. The person cared for should be able to depend on the competency of the caregiver and that this competency will be devoted to what is in the patient's best interests. At the same time, concern for the character of caregivers entails not forcing them to violate their consciences. When there is disagreement, there should be discussion and every effort should be made to resolve the matter in a way that reflects the patient's preferences without violating the caregiver's conscience.[23]

The concepts we have described that define the ethic of care can be used to mediate ethical disputes. This approach to resolving ethical issues is not principle- or theory-driven. Dilemmas are not resolved by an appeal to principles which are dominant. The approach focuses more on a relational ethic based on care as well as the guidance offered from using ethical principles. We all know that the nature of the questions posed influence the discussion and outcomes of a dilemma. The ethic of care encourages us to raise clarifying questions from a different perspective. The following example, although not a detailed narrative, will serve to illustrate the differences in the types and styles of the questions used in resolving the issue.

Imagine the following situation that involves a seventy-eight-year-old woman who has refused surgery. Evelyn has been given sufficient information, has considered the options, reflected on her values and still has refused surgery because she knows her problem isn't immediately life-threatening. She just wants help controlling the symptoms and staying comfortable. She says, "I want to concentrate on the quality of the time I have remaining." Evelyn has the capacity to decide and is competent. Questions about her competency have been raised, but by individuals who want to impose their wishes on her. The problem is in part how the dilemma is being described and the questions are being raised to support a particular view of the dilemma and its optional outcome. As a result, the central issue becomes whether or not to do the surgery even though she has objected. Ultimately, rather than focusing on what the patient wants, a plan to have her daughter sign the informed consent document is enacted and her daughter is asked and allowed to sign the form.

This travesty could have been avoided if the situation had been approached from a care-based or relationship-based point of view. The focus was on "Did we get the informed consent sheet signed?" Other questions could have reflected a different set of priorities that were relationship-based

23. Ibid.

and predicated on the ethic of care. "How do we show honor and respect to this woman who disagrees with us?" "How can we encourage communication between the caregivers and this patient and/or the patient and her daughter?" "Is what we are doing as caregivers respectful, compassionate, or kind?" The very nature of the questions and the rationale underlying these two points of view are different and influence the resolution in very different ways. The focus of the ethic of care in situations such as this is on enhancing relationships and keeping communication open. There is no problem getting people to talk when they agree in a situation; however, it is an art form to get people to communicate when they have strong differences of opinion and perhaps even differences in power and authority. Recognizing the importance and nature of a therapeutic relationship, which is a beneficence-in-trust relationship, is thinking that reflects the ethic of care. Such an ethic champions the importance of benefiting the person, based on her perception of what's good and her values. In this situation, a very different notion of what would be considered a right or good decision would be supported by the ethic of care.

One further essential component of therapeutic relationships and the ethic of care is the element of presence. Just what is presence and what is encompassed in presence? According to Pettigrew, presence involves coming alongside in suffering and the giving and sharing of oneself and one's skills. Presence means being there wholeheartedly and embraces a genuine willingness to be involved and available to the individual who is vulnerable and has needs. Presence also encompasses attentive and active listening — which is different from hearing. Listening involves hearing the things that are said as well as paying attention to what is not spoken, but communicated nonetheless. This kind of listening is done not only with the ear, but also with the heart. Presence includes being aware of the privilege of "ministering to" and "serving" the individual. Neither of these terms is very popular in today's health care literature, and there is often silence when these values are given a voice. Presence is by invitation, where the caregiver is extended the privilege of sharing in the person's circumstance, fear and pain. The caregiver should never assume the *right* to intervene based on one's role, title, or expertise. No matter what our profession, we know as healers that vulnerability invites vulnerability. Presence is a gift we share. Pettigrew believes that "the healing power of vulnerability comes as a result of the [caregiver's] willingness to be there in the midst of a helpless situation rather than focusing on saying or doing a particular thing."[24]

24. Jan Pettigrew, "The Ministry of Presence," In *Ethics: Critical Care Nursing Clinics of North America,* ed. D. Uustal (Philadelphia, PA: Saunders, 1990), 506.

Research shows that when presence is experienced, individuals report that a genuine encounter occurs and a positive relationship is created. "The person really was interested in me, and cared about my problems and reactions." People notice an inspiring quality to the relationship that is experienced in simple things such as the person's countenance, attentiveness, touch, and the tone of the person's voice. "She brought out the best in me and I was able to face the situation better. I gained courage, was more grounded, and able to think more clearly." Surprisingly most people indicate that the quality of the relationship is not directly related to time and that not a lot of time is needed to experience genuine presence. Time is not the essential component. Instead, what is essential is the investment and giving of one's whole self in that moment. People who have experienced presence describe it as meaningful and supportive, enduring and remembered. Presence can also be experienced even when people do not have direct, face-to-face contact, particularly as a follow-up to such contact. According to Pettigrew, "Spiritual presence can exist apart from physical presence."[25]

There are additional outcomes that are reported as a result of a relationship that involves presence. Literature and research indicate that people report that they experience an increased ability to cope and a decreased sense of vulnerability. They frequently report a decrease in the feelings of isolation, alienation, and suffering. In addition, individuals share that they experience a sense of caring, a genuine connection, and an increased sense of trust with their caregiver, a feeling of being "heard." As a result, they say that their situation feels less intimidating.

The core values inherent in the ethic of care are rarely consciously utilized as the basis for the goals of patient/client care or for written nursing and medical care plans. Utilizing these core values to guide caregivers in their planning and response would result in marked improvement in the continuity of a person's care, could help to continually improve service and care, and could enhance multidisciplinary, collaborative practice. Such core values could offer a framework for ethics committee proceedings and decision-making processes, and could help the members examine and rearticulate hospital policies. They could also help professionals to examine hospital policies and strive for care that is ethical, beneficence-centered, focused on a commitment to compassion, and justice-oriented. As a result, healing relationships would be enhanced.

25. Pettigrew, "Ministry of Presence," p. 508.

The "Invisibility" and Fragmentation of Care

Why is there such inattention to the ethic of care and such a complacent attitude about the importance of care and caring in relation to health, healing, and well-being? First, care does not command the same attention, response, and respect from the public or professionals as cure does. Could this be because the media dramatically celebrate medical cures and high-tech interventions, yet less frequently provide a forum to celebrate the inspiring stories of care and healing (even in the face of death) that change the quality of life for both patients and caregivers? It does not seem to be so glitzy or compelling to feature the essential caring behaviors of providing comfort, compassion, support, and presence. They seem pale and less important in comparison. In addition, care and caring are not generally regarded as "scientific" or technologically sophisticated — characteristics that we have been taught to value in our culture.

Even in the health care arena itself, caring behaviors are not rewarded or recognized in the same ways or with as much excitement as curative behaviors. New cures are rewarded by higher payoffs and more immediate reaction. Since there is a disproportionate amount of attention paid to curing modes as opposed to caring modes, there is more funding for cure than care research and more financial backing in health care for curing than for caring activities. To the extent that caring has been more valued by women than men, a traditionally male-oriented society has tended to underemphasize care.

The lack of attention paid to caring and its relative invisibility is not a problem unique to the United States or to this time in history. Care and caring are typically not noticed and valued until they are missing. They are more often associated with women's work and are unfortunately systematically devalued, under-recognized, and under-compensated. Care and caring are often regarded as "soft" concepts that focus on emotion and feeling. In 1859, Nightingale clearly recognized this problem when she declared in her book, *Notes on Nursing, What It Is and What It Is Not:* "So deep-rooted and universal is the conviction that to give medicine is to be doing something, or rather everything; and to give warmth, cleanliness etc., is to do nothing . . . the very elements of nursing are little understood."[26] Over a hundred years after Nightingale penned these words, we still find that the critical importance of care and caring (and even of nursing) is either not understood or misunderstood. She passionately argued for reform in health care to be based on human car-

---

26. Florence Nightingale, *Notes on Nursing: What It Is and What It Is Not* (London: Harrison & Sons, 1859), p. 6. Cited in *Notes on Nursing* (Philadelphia: Lippincott, 1992).

ing, healing, and health values and urged that the private work and world of women's caring, and the knowledge base and wisdom that are inherent in caring, should become a matter of public significance and political support. Watson draws inspiration from Nightingale and states that "this private ancient feminine wisdom, once brought to public and political consciousness, can become the foundation for societal and systemic reform, if not transformation." She continues by challenging nursing (and all health care professionals) "to recommit to a 'calling,' to engage in reform based on basic human caring-healing and health values."[27]

Caregivers everywhere complain that the consistent interference with "being able to care the way we were taught to care" because of inadequate staffing and time is what leads to burnout and compassion fatigue. The constant assault on the therapeutic relationship because of downsizing, rightsizing, and the frequent transferring of individuals to cover for inadequate staffing is a source of frustration and ethical conflict. Care and staffing based on patient needs, acuity, and safe care are radically different from staffing done with a skeleton crew driven by the pressure to focus on the "bottom line." Admittedly, we must be good stewards of our resources and should be cost conscious. However, when the continuity of a person's care and patient safety are compromised because of a zealous emphasis on the bottom line, on cost containment (which is not always cost effective), and on profit, real harm can and does result. Continuity of care and therapeutic healing relationships must be seen as vital to an individual's healing. We would see a totally different health care environment and outcomes if the ethic of care were truly embraced and practiced.

"Do what you value. Value what you do." Dr. Sid Simon, this author's mentor, shared these two brief messages often. They are simple life lessons, but they are not at all simplistic. It is a challenge to translate into behavior the things we *say* we value. As professionals, we say that we value caring. We talk about it as an ethic, an art and a science; however, merely defining and discussing it without translating it into action is anemic at best. We cognitively agree about its importance, but rarely hold professional caregivers formally accountable for care and caring in systematic ways. As Plato said, "A society cultivates whatever is honored there. It is important for us to know what we honor."[28]

27. Jean Watson, "Notes on Nursing: Guidelines for Caring Then and Now," in *Notes on Nursing* (Philadelphia: Lippincott, 1992), pp. 80, 81.

28. Plato, as translated and cited by L. D. Weatherhead, *Psychology, Religion and Healing* (London: Hodder & Stoughton, 1951), p. 112.

Do we really honor and value caring in our culture of health care? Or are there other business-oriented values and models that have trumped the importance of caring in therapeutic, healing relationships? Caring is still not seen as scientifically significant and continues to be "invisible" and devalued. A business-oriented model of care has overshadowed the beneficence-in-trust model based on the ethic of care and even the language we use in health care reflects this shift in values. Both the public and professionals are disquieted by the criticism that health "care" is delivered by people who are often perceived to be uncaring.

We need to reaffirm and restore care and caring as essential in therapeutic, healing relationships. Caring needs to become a valued, ethical cornerstone and needs to be experienced by patients, clients, and families in ways that are genuine and therapeutic. To treat a person only as a diseased body to be repaired without a caring attentiveness and presence is equivalent to psychological and spiritual abandonment. Caring is an ethic that links us together as members of the human community, and the absence of it denigrates both the caregiver and the patient. The significance of caring as an ethic and the importance of providing the best possible care must become central themes in health care and distinguish our progress and our healing environments in the twenty-first century.

PART IV

PROACTIVE PERSPECTIVES

Facing the Future

RICHARD A. SWENSON

Health care is changing at an unprecedented rate with no deceleration in sight. Never in history has the paradigm shifted so suddenly and so massively. What we had is gone; what will be is not yet here. We live in an awkward destabilized parenthesis.

Speculation on the future of health care and bioethics is in an inflationary phase. Having navigated ourselves off the map, we do not know what is around the next bend. Organizational guru Peter Drucker offers his own interesting perspective on our current debate: "The best way to predict the future is to create it." This is precisely what we must do. The uncertain time of the parenthesis represents a unique opportunity to reexamine health care and match it to the kingdom rather than the culture.

Guidelines and Principles for Thinking about the Future

Before offering specific predictions for our near-term medical future, there are several guidelines and principles to consider. Ignoring any of them could sabotage the reliability of our predictions and leave us ill-equipped to manage the cascading ethical issues that lie ahead.

Objectivity

In order to forecast the future accurately, it is important to be clinical and analytical. Objectivity must be carefully guarded as a fragile commodity. People

159

often lapse into emotional reactions against new developments. Such emotionalism, however, results in more irrationality than enlightenment. Emotions are best checked at the door. Anger, hostility, apathy, depression, despair, cynicism — such frustration will only mislead the futurist and often result in overreaction.

Trend Perceptivity

Trend perceptivity is not an intuitive skill. In order to see the flow of issues and events over time, a specific trend mindset must be developed. This requires the hard work of both lateral and longitudinal integration. Lateral (or horizontal) integration requires looking across the breadth of our contemporary spectrum and connecting everything. Longitudinal integration requires taking the laterally integrated whole, extending it backward and then finally extrapolating it into the future. Everything must remain connected. Only when full integration is achieved is it possible to assess accurately the direction of issues and events. Whether we like it or not, history steamrolls us with the whole, not with particles.

Cultural Context

All issues float in a context. It is important to understand this cultural context, for the direction of the future will be shaped and influenced by its values. In this light, it is helpful to note that Western culture is now wildly pluralistic. A few decades ago the Judeo-Christian ethic largely ruled the day and held sway. Today, however, that Judeo-Christian ethic is only one of many beliefs people are choosing from. Pluralism will continue to complicate the scene, making consensus (ethical and otherwise) increasingly difficult.

Neopaganism

Another dimension of the cultural context is the rapid slide of Western culture from Christian to post-Christian to neopagan — a naturalistic and/or materialistic worldview, often emphasizing a New Age spirituality. An examination of the popular media, especially movies and music, will suffice as evidence to secure the point. In bioethics, the single issue of partial birth abortion serves to make the neopagan case. This neopagan diagnosis is important to understand, because a neopagan culture will respond to medical and bioethical issues very differently than the previous Judeo-Christian world-view would have.

Realism

Those wishing to be accurate in the work of futurism must be prepared to see *what is,* rather than only seeing *what they wish to see.* Optimism reaps provable health benefits in life, but as a context for futurism it is seriously flawed. Pessimism likewise is flawed. The serious thinker must insist on realism. It is of utmost importance to follow the trend lines and allow them to take us where they are going, not where we *want* them to go.

Ultimately, the only thing that matters is whether the predictions are true. Neither optimism nor pessimism will determine that outcome. In the end, the fate of medicine and bioethics will not be determined by our view of things, but instead by the truth of things. We do not judge that truth — it always judges us.

Fallenness

Fallenness, said G. K. Chesterton, is "the one Christian belief that is empirically verifiable."[1] Even those not spiritually minded can detect a distortion in the cosmos that consistently thwarts perfection (often generically-labeled "unanticipated consequences"). Because our world is fallen, there is at least something wrong with everything. There is a flaw — a defect — inherent in everything on earth, and the presence of this flaw is universal. It is universal longitudinally across time and it is universal laterally across all human experience.

Accurately understanding fallenness gives the futurist a tremendous advantage in explaining why things go wrong. As an empiric model, it works. If we wish to understand and predict the future accurately, fallenness must be integrated into both our analysis and prognostications. If fallenness is ignored, we place ourselves at significant disadvantage in understanding our current world and our probable future.

It is worthwhile to differentiate between fallenness and pessimism, for

---

1. See John Ortberg, *The Life You've Always Wanted: Spiritual Disciplines for Ordinary People* (Grand Rapids, MI: Zondervan, 1997), p. 188.

they are often confused. *Pessimism* is a predilection to take the darkest view of things, to choose the gloomiest of all options, to see always the negative rather than the positive, to accept whatever interpretation is the most onerous, to believe that evil outweighs good.

Fallenness, on the other hand, simply acknowledges that things are inherently defective. It does not say that this defect always prevails (as pessimism often does). It doesn't refuse to acknowledge the beauty and happiness in the world (as pessimism generally does).

Pessimism maintains that the good *will not* win. Fallenness maintains that the good *might not* win. Pessimism maintains that nothing is ultimately promising; fallenness maintains that nothing is ultimately pure.

When we understand fallenness, then we understand why things so often end up flawed in mysterious ways. We understand why the many advantages we have in progress still do not completely offset the serious problems in the world system. We understand why, with each new "thing" added to the world stage, there is a downside that must be acknowledged, anticipated, and dealt with. We understand why Utopia did not show up as expected.

All future developments in medicine and bioethics will be tainted by a degree of fallenness. Every step we take opens up two new universes: The universe of all the positive that can result from a development, and the universe of all the negative that can result from this same development. One of the important roles of ethics is to decide where the dividing line is between these two universes and to guard against malignant effects of the fallenness.

Evil

As we move forward into a future of stunning medical developments, an important corollary to fallenness presents itself: evil. With new developments, in what way will we facilitate the purposes of evil? A guiding principle is not to give opportunity for evil. When evil has more power or more opportunity to express itself, the brakes must be applied. In Nobelist Isaac Bashevis Singer's story *Gimpel the Fool*, a rabbi reassures Gimpel by telling him: "It is written: Better to be a fool all your days than for one hour to be evil."[2] If we do not wish to contribute to evil, at times it might be necessary to forsake some betterment. This should not paralyze us — but it should make us careful.

2. Isaac Bashevis Singer, *Gimpel the Fool and Other Stories* (New York: Fawcett Crest, 1980), p. 13.

Convergence

A compelling formative ingredient in future developments is the much-discussed phenomenon of convergence. There are many definitions for convergence, but a functional definition is *everything coming together at the same time.* This phenomenon is seen in such areas as medicine, economics, technology, information, computers, Internet, media, travel, telecommunications, weaponry, population, and globalism. Each item listed plays a significant role in making our era different from all those that preceded it. When we factor in the interrelatedness of issues, the dimensions involved, and the speed of change, the word "unprecedented" comes to mind.

Unprecedented Age

Because of convergence, we live in a historically unprecedented time, a period of profound historical discontinuity, a historical disjuncture. We stand on a threshold that is new, exciting, and frightening at the same time. Yet seldom do people realize the historic when living through it. To be called to live in such a historically pregnant moment is a great privilege. It is an exciting era with much spiritual opportunity, even as it is also a frightening era with much spiritual danger.

Exponentiality

The math today is dramatically different. History has shifted to fast-forward. Many of the linear lines that described the past have disappeared, replaced with lines that slope upward exponentially. Graphing any number of contemporary social phenomena over time reveals such an exponential dynamic. Exponential change is rapid and disruptive. Such change can go from *little and slow* to *dramatic and fast* almost overnight. The significance of this is incalculable, yet our intuition is always behind the curve in understanding this phenomenon.

To illustrate how rapidly exponential numbers accumulate, fold a piece of paper in half forty times. How thick is the result? It would reach to the moon. If folded one hundred times, it would reach the far wall of the universe fifteen billion light years away.

Another example: The Pacific Ocean is sixty-three million square miles, so large that all the continents of the world could fit within it and still have room for another Asia. Additionally, the Pacific averages 14,000 feet deep. If we dried up the Pacific Ocean and then filled it again by doubling a single drop, how many doublings would be needed to refill the entire ocean? Eighty doublings. How full would it be on the 70th double? One-thousandth of the way full. With exponential change, most of the action happens on the backside of the curve. Someone once asked Ernest Hemingway how you go bankrupt. "First gradually," he said, "then suddenly."

As a result of exponentiality, we stand on the threshold of something very interesting. Developments will unfold suddenly. And we will consistently underestimate them.

Imagination

In the past, imagination had a credibility problem as science fiction writers over-imagined the future. Possibilities were embellished to histrionic levels. Now, however, imagination might well have a different problem: under-imagining what will happen. Imagination is no longer sufficient to assess the future, for it misleads us by underestimating rather than overestimating.

Interface

The temptation for analysts is to spend most of their time and effort examining the immediate interface between today and tomorrow. Most policy people tinker around the edges of our current paradigm. It is far more beneficial, however, to skip a generation or two of change and look ten years down the road. Such an approach is, in fact, the only way to adequately anticipate and plan.

Issues and Predictions concerning the Future of Health Care and Bioethics

Utilizing the above principles and guidelines, a very interesting future comes into view. It is filled with great promise yet also with a corresponding trepidation based on objectifiable concerns. Not every reader will agree with the following assessments, but they represent a good faith effort to anticipate the promises and threats of our collective future. There is no time to lose in preparing for these dramatic events. The United States serves as the central example here for illustrative purposes; but many of the trends are worldwide.

Costs

To understand where health care is going, watch the cost curve. There are other important factors to be sure, but cost is king. Economics dominate the present agenda and will dominate the future even more decisively. This single issue will largely rule the debate and ultimately collapse the paradigm.

Costs are inexorably increasing. Consider the situation in the United States. Total health care expenditures already are $1.2 trillion annually, and the Health Care Financing Administration (HCFA) predicts an increase to $2.2 trillion by the year 2008 (>$7,000 per capita). Even as a percentage of GDP, health care will go from 5 percent in 1960 to a predicted 16.2 percent in 2008.[3] There will be more and more people living longer and longer with more and more chronic diseases, taking more and more medications (that are ever more expensive), using more and more technology, with higher and higher expectations in a context of more and more lawyers. These trend pressures are all increasing, and the convergences are exponential. As long as the overall cost graph remains an exponential curve, the destabilization of medicine will continue, reactive change will persist unabated, and any future equilibrium will be postponed.

Managed Care

In the early 1990s, with seemingly uncontrollable costs, the apparent best hope for this runaway problem in the U.S. fell under the banner of managed care. That experiment helped decompress costs for a period. Today, however, the promise has largely soured. Managed care is itself besieged by a host of problems, not the least of which is — yes, it's true — runaway costs. The backlash is broad-based: patients complaining, shareholders suing, politicians hog-piling, and physicians counter-organizing. Managed care will continue to be a part of the reconfigured health care delivery system, although not as prominently as during the past decade.

3. Report titled "National Health Expenditures Projections: 1998-2008," prepared by The Office of the Actuary, Health Care Financing Administration. Available at http:// www.hcfa.gov/stats/NHE-Proj/proj1998/hilites.htm.

Employer-Based Health Care Delivery System

Most people in the U.S. (currently about 150 million) receive health care coverage through their employer. Only two countries in the world have such a system of health care financing. It is a system that has worked well for many decades but today is under great fiscal pressure. For example, from 1948 to 1990, the amount businesses paid for health care increased 15.6 percent per year — obviously an unsustainable cost escalation.

When the Clinton administration failed in its massive health care reform initiative, market forces stepped in without missing a beat. Managed care was the hoped-for messiah. These measures worked for several years, but short-term gains are not the same as long-term reform. Today, frightened people in every quadrant are looking at each other, asking: "What's next?" The answer: there is no next. Nothing can save the current paradigm.

As costs continue to rise, now unchecked even by managed care, the employer-based model of financing health care will be abandoned in its present form. This will happen, however, only when the Fortune 500 companies understand such a move to be viable from both a political and public-relations point of view. Employers will instead give defined contributions (as a part of the benefits package) equal to the amount of the insurance, and individual patients will purchase their own coverage. This will hopefully permit new regulatory openings for faith-based alliances. Those not included in the employer-defined contribution system will receive government vouchers or tax-deductibility for self-purchased insurance.

Uninsured

The problem of the uninsured seems refractory. Currently the number of uninsured stands at 44 million and is growing by about a million per year. Various predictions estimate it rising to 50-60 million. Sixty to eighty percent of uninsured are members of working families.[4] If we move in the direction of systemic health care delivery changes with a combination of government-employer-individual vouchers plus a democratization of health care delivery, the problem of the uninsured will be eased. If such change does not occur, however, this problem will continue as a painful social problem. This dilemma also extends to the intersecting problems of

4. Reuters Health News, "Rising Health Premiums Will Boost Ranks of the Uninsured," May 5, 2000.

the unemployed, the underemployed, the self-employed, and the small-business employed. Whose responsibility is this? What is the church's ethical response?

Democratization of Health Care Delivery

What Luther did for the priesthood of all believers in religion, the Internet will do for the democratization of all patients in health care. This new and radical democratization of medical information and health care delivery will result in patients becoming their own primary care providers. Although this new system will be awkward for some medical practitioners, it will ultimately dominate the landscape. Once people understand they do not have to negotiate with a monolithic system to check their cholesterol, they will flock to endorse the new system. Instead of paternalism, the new model will be one of both autonomy and accountability.

A simple illustration might help to clarify what this system will look like. Thirty years ago, physicians did not want diabetic patients to know their own blood sugar levels. Such knowledge would tempt diabetics to make bad decisions about manipulating their own insulin levels. The result, it was thought, would surely be subtherapeutic and possibly fatal. Over the years, attitudes shifted. Under the new paradigm, diabetics are encouraged to know absolutely everything about their disease. Thorough education is given; freedom is increasingly granted to make changes in diet, activity, and insulin; and frequent glucose testing is the norm for all diabetics. Today it is regarded as abnormal behavior if diabetics do not take control of their own disease management.

In the near future this same model of patient autonomy and accountability will extend across the entire medical spectrum.

- **Home Testing** for and monitoring of hundreds of conditions will become a widespread reality as patients buy over-the-counter kits in pharmacies or through the Internet (already three hundred such tests are available). People will own inexpensive diagnostic testing equipment, such as home-based lab testing machines, EKG equipment, etc.
- **Lab and X-Ray** by self-referral will increasingly be the norm. For-profit freestanding lab and X-Ray stations will arise across the nation, allowing people to stop by without an appointment or prescription and obtain virtually any test they desire. Already such stations exist in selected market locations testing the profitability of the concept. In addition to a

complete offering of lab profiles, such stations offer EKGs, CAT scans of the coronary vessels, and even rapid CAT scans of the total body.[5]

- **Home Computers** dedicated for health care, or even wireless palm computers, will be standard and possibly free (if advertising, perhaps mostly health related, is allowed). Already, for example, Ford Motor Company has passed out thousands of free computers to their workers. Ford wants its workers to be computer literate, and management determined that perhaps the best way to accomplish such learning was to put free computers in workers' homes where learning could happen off company time. Insurance companies and health care alliances will come to a similar awareness and offer free or low-cost health care computers to patients.

- **Smart Cards** will allow patients to control their entire medical record. Smart cards have caught on in Europe and Asia but have penetrated more slowly in the U.S. Once the technology is widespread and people see what options are available to them, however, the shift will come quickly. All practitioners will use voice-activated electronic patient records, and patients will not only hear the dictation but will fully know what is in their charts. They will have all this information loaded on their smart cards. Once this trend begins, it will be impossible to stop. Patients will reason: "It is my body, my health, my money, and my future. Why shouldn't I control my medical record?"

- **Internet** shopping will allow patients to access a broad range of health care services from their own home: buying insurance, ordering prescriptions, accessing complete health information in simplified formats including entire textbooks, journals, and the Physicians' Desk Reference. Health-related web sites already number between 10,000 and 100,000. The majority will not survive the inevitable consolidation, but their presence will nevertheless transform the landscape.

- **E-mail** will replace telephone tag. Lab, X-Ray, biopsy results, patient education materials, informed consent documents and insurance papers will be transferred electronically to the patient's designated health e-mail address in encoded and secure formats. Interestingly, the Costa Rican government recently announced it would sponsor free e-mail for everyone in the country.[6]

5. Julie Appleby, "Want a CAT Scan? Step Right Up Medical Labs Offer Tests on Demand for Those Willing to Pay," *USA Today,* May 25, 2000.

6. Elliot Blair Smith, "Costa Rica Online with Free E-mail for All," *USA Today,* June 1, 2000.

- **Medications** will be more available in over-the-counter (OTC) formats. When histamine H$_2$ blockers — a form of ulcer medication — first became available, they were so tightly regulated that only gastroenterologists could prescribe them. Today, these same medications can be bought cheaply OTC. Restrictions on many other prescription medicines have similarly been relaxed. Current efforts are underway to allow OTC status for birth control pills, antihypertensives and cholesterol-lowering agents.

The ethical dilemma for physicians, patients, and policy makers alike is whether to resist or embrace such democratization. Should it be denounced as a threat to the doctor-patient relationship? Will patient pseudo-sophistication lead to dangerous decision making? Or should it be embraced as an inevitable and cost-saving transition into a new era of patient empowerment? Actually, the opinion of the current health care paradigm makes little difference. If the trend pressures are strong enough (and they are), the marketplace will make the move with or without the permission of the reigning medical establishment.

Aging Demographic

At the time of Christ, average life expectancy was 21 years. By 1900, it was 47 years. Currently, it is 76 years . . . and still rising. As life expectancy rises to 100 and then possibly 120 years, what will society do with the increasing numbers of the aged? Currently there are more than 33 million seniors over 65; by 2025, these numbers will swell to 62 million.[7] Our nation has not solved this difficult social problem even today. What then will it be like when our grandmothers are themselves caring for *their* grandmothers? No socially or fiscally stable answers have yet presented themselves.

Might we increasingly become as the Cumaean sibyl in Virgil's *Aeneid,* who asked to live forever but forgot to request perpetual youth? She became more and more shriveled until finally her servants had to carry her around in a basket. The children would call into the basket, "Sibyl, what is it you wish?" "To die," came back the response.

---

7. "America's Seniors and Medicare: Challenges for Today and Tomorrow," 29 Feb. 2000, National Economic Council/Domestic Policy Council: http://www.whitehouse.gov/WH/New/html/Medicare2000/.

Medicare

The increasing numbers of elderly are already straining Medicare, which covers 39 million people at an annual cost of $220 billion. What will happen when Medicare experiences the greatly increased pressures as baby boomers (the 77 million Americans born between 1946 and 1964) begin retiring around 2010? Even though our elderly are healthier and wealthier than they have ever been, at the same time longevity gains are also leading to a higher incidence of chronic illness and associated costs. In 1995, for the first time ever, more people died of chronic diseases than acute conditions. Eventually utilitarian motives, costs considerations, and a misunderstood "compassion" will result in sweeping new laws permitting the active, premature ending of life with both assisted suicide and euthanasia.

Abortion

The painful and refractory issue of abortion will become medical rather than surgical. The French abortion pill RU 486 or other substitutes (such as a combination of methotrexate and misoprostol) will result in a high rate of successful medical abortions. As a result, patient and clinician anonymity will leave few options for protest. The only battle will then be the struggle a woman will have with herself, as she faces the mirror with a pill in one hand and a glass of water in the other. At this point, the advocacy for the infant will lie only within the conscience of the mother — ironically, where one would have thought it the safest.

The partial birth abortion debate is a cultural marker on this tragic issue. If even the battle over partial birth abortion cannot be successfully won, the larger challenge of overturning Roe vs. Wade seems almost hopeless.

Prenatal Testing

Prenatal testing, already mandated in many states for specific conditions and enforced through threat of liability in other states, will expand to include hundreds of conditions as testing technology becomes available. Prenatal DNA testing could be performed via chorionic villus sampling at 9-11 weeks gestation. An alternative would be via preimplantation genetic diagnosis (PGD), where the mother's egg and father's sperm are mixed in a Petri dish and the resultant fertilized eggs are subjected to intense DNA analysis. Any physicians failing (or refusing) to offer such tests will find themselves potentially subject to liability involving medical/suffering costs for a "defective" child thus neglectfully born.

Human Cloning

Successful human cloning is a certainty, most likely within the decade. The initial ethical protests will be quickly drowned out. Some reports indicate that human cloning has already happened in experimental stages with the early embryos being destroyed rather than implanted. Will we have women giving birth to themselves? Will there be resurrections from the dead?

The French cult Raelian Movement has formed a Bahamas-based company called CLONAID, the world's first commercial human cloning service. For "as low as $200,000," CLONAID will provide a cloned child to anyone wishing it, including "parents with fertility problems or homosexual couples." If a man should die early in life, "imagine the joy of a widow raising a child looking like her beloved deceased husband." What a twist on the Oedipal complex. This cult believes that life on earth was created scientifically in laboratories by extraterrestrials whose name (Elohim) is found in the Hebrew Bible and was mistranslated by the word "God." It also claims that Jesus' resurrection was a cloning performed by the Elohim. For a $50,000 fee, they will provide INSURACLONE, the sampling and storage of cells from a loved one to be cloned later if the beloved person dies unexpectedly. Farfetched? They expect one million cloning customers.[8]

Genomics

The mapping of the human genome will leave many questions and challenges. Gene manipulation and therapy, although initially disappointing, will eventually become widespread. Once technically successful, genetic manipulation will revolutionize medical therapeutics.

Some form of eugenics with designer babies is inevitable. Genetic structures will be placed within the genome of an embryo prior to being implanted into the mother's womb. After genetic screening, only an optimal embryo will be allowed to exist.

Stem cell manipulation using pre-differentiated cells will lead to organs

---

8. "Rael Creates the First Human Cloning Company," http://www.clonaid.com.

for transplant on demand, among other applications. *Germ cell manipulation* has the potential to change the genetic pool of the entire human species, enhancing human characteristics that will then be passed on. Such germline cell therapy is widely opposed. The question is: When we develop the technology, will we still feel the same? Will we be able to control ourselves? Will we be able to control others? If not allowed in the United States, this effort will easily go offshore.

Reproduction and Sex

Reproductive options will be limited first only by morality and the imagination, and later only by the imagination. When all the modern variables are considered (source of sperm and egg, site of fertilization, and whose womb is impregnated), there are now over twenty-five different ways to create a baby. It is reasonable to predict a near-total dissociation between recreational sex and reproductive sex. Cybersex will also be a widespread, disturbing, addicting reality. "Anything that can be used for sex," someone once said, "will be used for sex."

Technology

Technological advances will stun even the cynics. "Infinite energy" is predicted by the middle of the century via nuclear fusion. If such a promise succeeds, the applications will be overwhelming. The question must be asked: with the power of infinite energy, what level of mischief will fallenness be capable of? Nanotechnology will manipulate atom upon atom, building any product imaginable at the elemental level. The possibilities for biotechnology as well as global manufacturing will turn the industrialized world on its head. IBM has now come up with a computer that performs twelve trillion calculations per second.[9] While this is still one thousand times slower than the human brain, eventually computers and robotics will equal human abilities and then surpass them. Posthuman scenarios are being seriously predicted, where humans merge with robotic species. Sun Microsystems cofounder Bill Joy, writing in *Wired* magazine, is candid in his warning of "knowledge-enabled mass destruction" and "the perfection of extreme evil, an evil whose possibil-

9. D. Ian Hopper, "IBM Announces New Supercomputer," The Associated Press, June 28, 2000.

ity spreads well beyond that which weapons of mass destruction bequeathed to the nation-states, on to a surprising and terrible empowerment of extreme individuals."[10]

As dramatic as the above projections might seem, the sea change ahead has still not been fully defined. Scores of additional changes are coming, such as the following:

- Telemedicine will become commonplace, particularly in underserved areas.
- Computers will make, or assist in making, many hospital decisions, including antibiotic choices, EKG interpretations, and even calculating mortality predictions on ICU patients, helping ethics committees with "pull the plug" decisions.
- Alternative medicine is "in," and gaining mainstream acceptance. It will continue to grow, as disenchanted patients insist on wide options.
- A wide variety of health practitioners will open shop. Pharmacists will prescribe directly.
- New "latent diseases" will wait in the wings for the opportune time to manifest themselves.
- Emerging infectious diseases will become serious threats, due partly to dramatic mobility patterns (ten million people per day cross international boundaries). Pathogens that we have not yet heard of will dominate our attention in the coming years.

In the world newly upon us, it is essential that the Judeo-Christian worldview have a place at the table in the national debate. We must first clearly understand the future we face, with issues prioritized into the most urgent and the most important. With the most urgent and important issues identified, conferences and panels could then do the difficult work of forging consensus-based ethical policy statements. Once these statements are formed, they can be used to attempt to influence the broader culture. *But even if we fail in this larger goal, we must succeed at least in living by these standards ourselves regardless of the cost.*

Given this remarkable era, one thing is clear: passivity is not an option. Can we take up this challenge and create the future of ethical health care for the sake of the kingdom? The culture, the church, and our children await our answer.

10. Bill Joy, "Why the Future Doesn't Need Us," *Wired*, April 2000. The entire article is available at http://www.wired.com/wired/archive/8.04/joy.html, pp. 1-11.

Media and Public Policy Challenges

HELEN ALVARE

Our world is exploding with information that bombards us daily via a number of media: television, newspapers, magazines, radio, the Internet, and so on. Influencing the course of society and public policy requires being able to communicate one's position attractively, practically, and clearly in a way that will stand out amidst the clamor of competing ideas. In this chapter, we will explore how to accomplish that goal in order to make a difference in the public arena.

A Matter of Vision

We must be able to convey to listeners a vision of a world that they would like to live in, and convince them that our position on a certain issue is an essential part of that world. We need to convey to them a vision of a culture that resonates with them on a deep level — a vision of a human person that they want to be on their best day — in order to motivate them to reach for their better self in a better society. And we do this through showing them how godly perspectives on important issues help to create that kind of person and society.

For instance, when talking about abortion, we would call people to a respect for life. We would not articulate a position in such narrow terms as "choosing" to be pregnant or "wanting" the child (based on the child's ability or disability, gender, or the situation in which a conception took place). Once we have articulated the vision, we would say, "Can't you see? Abortion doesn't fit in that picture. It will set up a situation in which we cannot reach the beautiful society we seek."

The key to successful media and public policy advocacy is shown in public policy campaigns that have worked before, such as civil rights. At its height, the campaign held up civil rights before us as a picture of society where people lived in harmony, no matter their race. People said, "We are not there yet, but we want to be. We want laws on the books in our country that say that we are a people who do not discriminate based on the color of another person's skin." We reached for the passage of the Civil Rights Act, even though we weren't as good as that law said we ought to be.

This kind of positivity in public policy campaigns and advertising is not at all unusual. Remember the Coca-Cola ad that sang, "I want to teach the world to sing"? It showed people holding hands around the world — people of every age and race in harmony — and somehow Coke was associating itself with this vision of a world. They said, if you associate yourself with our product, you're associating yourself with this image we know you want. We can learn much from this, because unless we articulate the positive vision first, before we critique or ask questions, then we will be perceived as anti-progress. We live in a world where what is new equals progress, and therefore those that express misgivings about what is new are considered to be expressing misgivings about progress. If that's our approach, then we lose our credibility before we have a chance to be heard.

In a country like the United States, if we are not for progress, we are not perceived to be a part of what this country stands for, at its best. So we have to be for real human progress, even if it is an alternative view of what progress entails. That which we positively advocate has to be articulated first, before any questions are raised about the opposition's perspective. Otherwise, we will put ourselves firmly in the camp of the "negative retro," where we will not even be part of the conversation about these cutting-edge technologies.

Look at organizations such as Celera Corporation and Genentech, who are promoting new technologies. They already seem to have a firm grasp of the word "progress." They have been able to paint a vision of a world where people are healthier on a daily basis. It is a world where suffering is reduced, our loved ones have a longer and better life, and some of the most common diseases can be eradicated. Of course, the money never seems to be used that way, even though we've had decades to do it. But they always promise that vision. Such organizations have practically aligned themselves symbiotically with the word "progress." That is why we have to put our own vision of progress before the public.

Challenges to Overcome

What are the primary challenges to the successful formulation and communication of our vision of progress in order to engage public policy and the media? The first challenge is to break through the knee-jerk reaction that what is new equals progress. We need to paint a more precise picture of progress. For the secular world, we should paint a picture in its language involving such things as: support for those who are most vulnerable, equal respect for every human being, egalitarianism and/or equal rights. However, within religious communities we can be more precise. For instance, in the Catholic community we would use the phrase *the option for the vulnerable*. We would say the worldview we are advocating is a world where we don't just treat everybody equally, but go out of our way for those who need us more. That's human progress. In other Christian communities, we might use the language of *loving as Jesus loved*. What does it look like to love as Jesus loved? It is to be the perfect human, and the model for the best authentic human progress is Jesus! So when we are painting a picture of what we mean by "the beautiful vision" and "human progress," we are talking about a world where people love as Jesus loved. Jesus exemplifies human progress. That was exactly the demonstration that His life gave, and that should be the framework in which we evaluate any new development.

The second challenge involves effective communication. We need people who can explain the stakes and ethical concerns to the public in pressured media, electoral, and legislative situations, and who can also explain what ought to be done. Explaining complex medical inventions and scientific developments in language that lay people can understand can be quite difficult. This difficulty is compounded by the dearth of public figures who are well informed and deeply concerned about potential threats to human life and dignity posed by emerging biotechnologies.

In this author's experience, some of the best-intentioned members of Congress are completely uninformed. They often are in the dark about the issues in view here. They depend on their aides to do their research and tell them what they need to know. They simply do not have the time, in many cases, to become knowledgeable, let alone to go the extra step of understanding the ethical problems of these new technologies. But we need them badly, if anyone is to stand up to large bodies of people such as industries and groups focused on specific diseases, who are knowledgeable, talented, and willing to promote a new technique if it benefits human health at all.

In order to overcome this challenge, we need to be prepared to inform our Congressional advocates as clearly and quickly as possible. In order to do this, the approach followed by the U.S. Conference of Catholic Bishops is instructive. We should devote significant time to being in contact with the scientists, the ethicists, and the doctors, training them to be intermediaries who translate difficult concepts into lay people's language. We can put their information onto fact sheets which they can use to tutor legislative leadership, or at least to get the ear of a faithful aide who is sympathetic. We can go over and over and over each new development with them in order that they can explain it as simply as possible. Language is important, and sound bytes are important. Everybody makes fun of that concept, but it works. It's analogous to how much makeup people have to slap on their faces to look natural before they get in front of a camera. In the same way, we have to play with language until we come up with a few effective, understandable phrases. This appears to be an artificial process, but it actually is very appealing to people and is an effective form of communication, so it is important to have people who are adept at it.

One final suggestion under this challenge is that it is not wise to be identified with standard pro-life groups when publicly addressing the ethical problems with a new technology. Such groups are commonly seen as "no" groups. No assisted suicide. No abortion. No capital punishment. We are the No people and we get a negative reputation out there because of this. No progress for women. No feminism. No enlightened thinking. No intellectuals. This is the caricature of pro-lifers. Public polls suggest that the most persuasive people the public can hear from are medical personnel, rather than the pro-life organizations.

Admittedly, there are some talented organizations out there, and it is a shame not to take advantage of their resources to help us get organized. But even the best pro-life organizations do better to stay, in many cases, behind the scenes. It is the churches who need to get out and speak and make their moral contributions to the public debate. If pro-life groups are going to get involved, they need to organize the experts and stay out of the picture. We are not fighting to receive credit, but to achieve a good result with creditable spokespeople using good means.

A third challenge, which applies particularly to the stem cell arena, genetic engineering, and new reproductive technologies, is the way the abortion debate is getting all mixed up with these questions. Pro-abortion groups are so desirous to minimize the value of nascent human life that they immediately publicize that to take a position that is protective of nascent human life is to be antifeminist and anti-choice. Recently, in support of stem cell research and trafficking in human body parts, pro-abortion groups have distributed press releases championing scientific research and the integrity and freedom of scientific researchers. Ironically, scientific researchers, in the exercise of this integrity and freedom, have conducted research that thoroughly documents the relationship between abortion and later psychiatric problems and between abortion and breast cancer. It is the results of this very exercise of their integrity and freedom that seem to argue most forcefully against the integrity of allowing people the freedom to take the lives of the unborn. Yet, the importance of moral limits — of limiting certain "freedoms" in order to preserve true freedom — is commonly overlooked.

One effective way that powerful and wealthy organizations like Planned Parenthood have found to weaken efforts to limit new biotech research constructively is to link such efforts to the movement to stop abortion. Because of this link, certain people who would otherwise listen to reasonable arguments for ethical research guidelines and who would resonate with our vision of the world are biased in advance. These organizations say, "Don't listen to those people. Those are the pro-life fanatics, those are people who think the embryo is more valuable than women." It is hard to get ahead of the pro-abortion groups on this because they are so powerful in the media. Before people get polarized into pro-life and pro-choice groups on some of the new technologies, we must find a way to reach people concerning the entire range of issues that are involved.

It seems that the most effective technique is honesty and straightforwardness, not only in presenting our own arguments, but also in clarifying the often convoluted statements of the other side. For instance, in a debate, we would do well to let our opponents bore the audience to death by stringing together a million phrases so that we can't get a word in edgewise, until the host of the show finally cuts them short. Then, instead of only making our own arguments, we should first point to theirs and restate them more clearly, but within our own framework. We might say, "now you see, what Planned Parenthood has just said is that there ought to be no limit on stripping stem cells out of embryos. We're really talking here about doing things to embryos on a mass level in a lab, which would horrify us to hear anybody suggest as an option for an embryo in a mother's womb. Supposedly these embryos are not anything because they are in a lab. But they are still human. Planned Parenthood is essentially supporting human farming." We should simply point to what they have said — rephrasing for clarification — and let it sink in.

Another challenge that we face is that disease groups — Parkinson's, the American Cancer Society, the American Heart Association — are coming out more and more in favor of any new technology that looks promising for their members despite the human cost. They're really not stopping to consider all that is at stake. They do not acknowledge the moral debate in their public statements, let alone seem morally troubled. When they bring in celebrities to Capitol Hill, like Christopher Reeve (now a quadriplegic) for stem cell research and Michael J. Fox for Parkinson's research, they are totally on the bandwagon.

In responding to the stance of these groups, we'll get nowhere if we don't begin by affirming their aspirations and acknowledging the suffering of their members, because that is what is driving them. In fact, when the NIH was talking about stem cell research, one of the members of the panel that was hearing testimonies forgot to turn off his microphone and urged the man next to him to find out all the diseases affecting the families of members of Congress so that they could appeal to them emotionally on particular issues. And that's exactly what they did.

People are very emotionally moved by the prospect of weakening or wiping out some of our most debilitating diseases. In order to be effective and be heard, we have to affirm their aspirations and sympathize with what they hope to achieve, while condemning means that further demean human life. For instance, we need to point out that there will never be respect for people with illness if we achieve their cures by disrespecting other people. The price of these cures cannot be our own humanity.

Another approach is to label publicly the use of celebrities and the petitioning of congressional families as cheap emotional tricks. One of the most important ways to do this is to highlight the differences between the promises for these new therapies and the actual test results. The tests of stem cells and fetal tissue have not come close to yielding the results promised in order to justify the research. When we notice that the initial speculation about proposed research gets huge media coverage but test failures get almost none, we should make sure we get out the news about the test failures.

Another important tactic is to point out alternative techniques. Fads and money chasing are as common in medicine as in other arenas of society. In this stem cell research craze, embryonic stem cells are the hot research property. This is not because adult stem cells aren't as promising as embryonic ones, but because embryonic research is a fad. In the same vein, the chase for certain research money takes on a life of its own, until it seems unstoppable. These "facts behind the facts" are not expressed in any public forum, and it is important for us to point them out, so that people are getting the whole picture.

Moving Forward

The challenges to drawing people toward a better vision of human progress are great, but it is possible to overcome them. Needed strategies require people who are willing to bridge the gap between the experts and the public policy and media figures. The experts are powerful, and so are public policy and media figures. Remember that both must be dealt with when crafting public policy.

It is also vital that we learn communication — which is a skill, not a deep body of knowledge. Communication is vitally important, and people who *can* do it well *ought* to do it actively. Those who are capable of bridging the gaps, experts who are capable of articulating, media people who are capable of understanding what the experts are discovering — all these must jump into this void and speak out. Otherwise, the talk shows and the blaring headlines are going to form people's moral conscience on today's bioethical challenges.

We are still at the beginning of the struggle. Where people have been exposed to the ideas of these new technologies, there is a lot of insecurity about the future. For this very reason, now is the time to jump in. The unease in the public creates an opportunity for us to transform their insecurity to informed support for beneficial uses of these technologies.

Ethical Challenges

C. BEN MITCHELL

By any method of reckoning, we have entered an age of nearly unbridled bio-technological expansion. Futurists almost universally claim that the twenty-first century will be what Jeremy Rifkin has called "The Biotech Century."[1] Vanderbilt University professor of business management Richard Oliver has announced that "The Bio-materials Age will complete the triumph of economics over politics, which was begun in the Information Age. It will unleash forces stronger than nationalism and more powerful than the combined armies of the world."[2]

The list of technologies is daunting; and, to coin a word, Oliver's characterization of this new age sounds extraordinarily "technopian":

- Creation of life in a lab.
- Predetermination of the sex of children and their genetic makeup.
- Pharmagenomics, which directs and tailors drugs to individual genotypes.
- The ability to "program" out of human genes the propensities to contract various diseases and illnesses.
- Genetically-derived therapies for the prevention and cure of most cancers, heart disease, AIDS, and other diseases, including new strains of vaccine-resistant ones such as malaria.

1. Jeremy Rifkin, *The Biotech Century: Harnessing the Gene and Remaking the World* (New York: Jeremy P. Tarcher/Putnam Publishers, 1998).
2. Richard W. Oliver, *The Coming Biotech Age: The Business of Bio-Materials* (New York: McGraw-Hill, 2000).

- Repair of damaged brain cells and spinal cords.
- Production of proteins that fight infections or treat problems such as growth deficiency.
- Mass production in a lab of at least six U.S. Federal Drug Administration (FDA)–approved monoclonal antibodies, which, when injected into a patient, hone in on the antigens that populate the surface of cancer cells.
- The ability to clone, or duplicate, mammals including humans.
- Control of aging and obesity.
- Animals that grow replacement organs for the fifty percent of humans who would die before getting a transplant organ from a human donor.
- Inexpensive "transgenic" vegetables that will produce vaccines capable of inoculating the world's poor against diseases that have ravaged them for centuries.
- A tree that will grow in a few years instead of fifty or a hundred, fundamentally changing the economics of everything wooden.
- A natural plant that will produce a substitute for the raw materials in plastic, potentially impacting the entire oil and petro-chemical industries.
- The world's strongest fiber and the world's most powerful adhesive produced by insect and animal "factories."
- A biological, protein-based computer thousands of times faster than today's fastest.
- Bio-electronic noses, tongues, ears, and heads to test industrial and consumer goods and provide new levels of real-time health care assessment.
- Bio-synthetic skin, blood, and bone, as well as the "precursor" human master cell that can be directed to grow new bone and cartilage.
- New materials for products and packaging that repair themselves and adapt to the environment.
- New materials that swell and flex like muscles to replace human muscle and machine power in factories.
- New materials that repel any ink, paint, or stain.
- New materials that shape and reshape themselves for a huge variety of industrial, consumer, and health care applications.
- New energy sources that are efficient, pollution-free, and almost cost-free.
- New paints that capture and store the energy of the Sun in cold weather, and repel its heat in hot weather, reducing energy costs and pollution associated with heating and cooling.

- A "smart mouse" that points the way to eliminating aging in humans.[3]

Clearly, the future may bring great benefits from biotechnologies such as genetic engineering, cloning, cybernetics, nanotechnologies, and a litany of neologisms yet to be invented; but the future may also portend human tragedy, a loss of human dignity, and a world increasingly hostile to concerns which transcend the purely materialist world of contemporary scientific research.

Are Christians even aware of these issues? Certainly some are. Does the church have anything to say about biotechnology? If so, what? If not, why not? Can we afford not to speak to these issues? Can we afford to mis-speak on these issues? These are sober questions for Christians who are witnesses to the dawn of the biotech age.

The Ethical Challenges Ahead

There are several issues which ought to cause us all to lie awake at night. They are issues which demand our most careful attention. And they are matters which will require a multidisciplinary collaboration if we hope to get a public hearing.

What Does It Mean to Be Human?

Increasingly we face the question, "What does it mean to be human?" In her volume, *How We Became Posthuman*, Katherine Hayles argues that mortal human beings are rapidly becoming an endangered species.[4] And if even a portion of Hans Moravec's vision of the future is realized, then human beings will have to fight for their own survival, but with an unlikely enemy. Says Moravec,

> Biological species almost never survive encounters with superior competitors. Ten million years ago, South and North America were separated by a sunken Panama isthmus. South America, like Australia today, was populated by marsupial mammals, including pouched equivalents of rats, deers [sic], and tigers. When the isthmus connecting North and South America rose, it took only a few thousand years for the northern placental species,

3. Oliver, *The Coming Biotech Age*, pp. 36-37.
4. H. Katherine Hayles, *How We Became Posthuman: Virtual Bodies in Cybernetics, Literature, and Informatics* (Chicago: University of Chicago Press, 1999).

with slightly more effective metabolisms and reproductive nervous systems, to displace and eliminate almost all the southern marsupials. In a completely free marketplace, superior robots would surely affect humans as North American placentals affected South American marsupials (and as humans have affected countless species). Robotic industries would compete vigorously among themselves for matter, energy, and space, incidentally driving their price beyond human reach. Unable to afford the necessities of life, biological humans would be squeezed out of existence.[5]

Notice that "biological humans" would cease to exist. So what kind of "humans" would survive? Here is what Moravec suggests.

> Humans can be enhanced by both biological and hard robotic technologies. Such present-day examples as hormonal and genetic tuning of body growth and function, pacemakers, artificial hearts, powered artificial limbs, hearing aids, and night-vision devices are faint hints of future possibilities. *Mind Children* speculated on ways to preserve a person while replacing every part of body and brain with superior artificial substitutes. A biological human . . . could grow into something seriously dangerous once transformed into an unbounded superintelligent robot.[6]

Lest we take these as the musings of a lunatic, we should note that Moravec is founder of the world's largest robotics program at Carnegie Mellon University. He is not unintelligent. Even if he were, a lunatic with the world's largest erector set would be a formidable power. Bill Joy, cofounder and chief scientist of Sun Microsystems, does not think these are crank ideas. He writes of his own concern about the ethical challenges ahead in his article, "Why the Future Doesn't Need Us."[7] And Joy does not discount the prognostications of Moravec or Ray Kurzweil at all.[8] In fact, he laments the fact that they might actually be right.

We must reestablish what, exactly, it means to be human. If being human is all about the brain, then supercomputers might be able to contain all the information in the brain. To the contrary, being human is all about having a mother. That is to say, according to the biblical witness, being human means being the offspring of human parents. Furthermore, beings which are

5. Hans Moravec, *Robot: Mere Machine to Transcendent Mind* (New York: Oxford University Press, 1999), p. 134.

6. Moravec, *Robot,* pp. 142-143.

7. Bill Joy, "Why the Future Doesn't Need Us," *WIRED* (April 2000): 238-262.

8. Cf. Ray Kurzweil, *The Age of Spiritual Machines: When Computers Exceed Human Intelligence* (New York: Penguin Books, 1999).

human are not so because they possess certain functional capacities like reason, volition, and self-awareness. Each of these functional capacities can be gained or lost. Humanness is neither gained nor lost; it either is or it is not. Human beings either are imagers of God or they are not human beings. Imagers of God either are human beings or they are not imagers of God. The mistake some of our systematic theologians have made is to unpack the *imago Dei* in terms of functional capacities. This is doubly deadly. First, it is contrary to revelation. The passages that speak to the image of God (e.g., Genesis 1:27; 5:1; 9:6) never divide the *imago Dei* into constituent parts. It is an ontological category rather than a category of functions. Second, as soon as one explicates a list of functions, capacities, or activities which are *sine qua non* to humanness, one capitulates to those who say that some humans do not have lives worth living. We have been down this hellish road before.[9]

The implications of this question are huge. They span nearly every biotechnology, including cybernetics and transgenics. We need to work on the question of what it means to be human. In sum, we need a new theological anthropology which takes into account some of the questions being posed in this new century.

What Does It Mean to Be a "Good" Human?

This is not a question about personal ethics, but about eugenics. The completion of the map of the human genome only brings closer the possibility of using this potentially wonderful technology as a weapon against the genetically undesirable and as a greenhouse for the genetically desirable. An earlier eugenics movement in the United States took the shape of the "fitter family" contests in the nation's heartland. In these contests, prizes were awarded to the families with the "best genes." Best heredity was measured by purest lineage, heartiest stock, and fewest disabilities, mental or physical. One contest brochure read: "The time has come when the science of human husbandry must be developed, based on principles now followed by scientific agriculture, if the better elements of our civilization are to dominate or even survive."[10] The 1924 Kansas Free Fair awarded a Gover-

9. See my "Of Euphemisms and Euthanasia: The Language Games of the Nazi Doctors and Some Implications for the Modern Euthanasia Movement," *Omega: Journal of Death and Dying* 40 (1999-2000): 255-264.

10. Cited in Daniel J. Kevles, *In the Name of Eugenics: Genetics and the Uses of Human Heredity* (Berkeley, CA: University of California Press, 1985), p. 62. For additional insight into this enigmatic period of American life, see Mark H. Haller, *Eugenics: Heredi-*

nor's Fitter Family Trophy, and the Capper Medal went to "Grade A Individuals."[11]

In other words, we practice eugenics for our livestock, why not for our children? With the human genome fully mapped, we are closer than ever to creating "better humans through biology." In fact, it is already happening. In 1993, a *New York Times* article reported that eleven percent of Americans would abort a fetus whose genome was predisposed to obesity. About four out of five said they would abort a fetus that would grow up with a disability. And forty-three percent of respondents to a March of Dimes poll said they would engage in genetic engineering simply to enhance their children's looks or intelligence.[12] In 1994, Singapore rewarded college graduates who produced children with a greater array of social benefits than were granted to nongraduates who produced children. And in 2000, "optimal" college women at universities across America were being solicited for their eggs to the tune of $80,000 per donation. This brings us to our next challenge.

Shall We Commodify the Human Body?

Obviously, many items are commodified in contemporary society. There is even a "commodities exchange." But commodification becomes problematic when it becomes a universal practice. By commodification I mean applying economic modes of valuation to items which have been traditionally the objects of noneconomic modes of valuation. Commodification is based upon two assumptions of market value: (1) "that there exists some scale into which every value inhering in a good can be translated" and (2) "that this scale is money."[13]

---

tarian *Attitudes in American Thought* (New Brunswick, NJ: Rutgers University Press, 1963); Troy Duster, *Back Door to Eugenics* (New York: Routledge, 1990); Diane B. Paul, *Controlling Human Heredity: 1865 to the Present* (Atlantic Highlands, NJ: Humanities Press, 1995); Edward J. Larson, *Sex, Race, and Science: Eugenics in the Deep South* (Baltimore: Johns Hopkins University Press, 1995); Diane B. Paul, *The Politics of Heredity: Essays on Eugenics, Biomedicine, and the Nature-Nurture Debate* (Albany, NY: State University of New York Press, 1998); Martin S. Pernick, *The Black Stork: Eugenics and the Death of 'Defective' Babies in American Medicine and Motion Pictures Since 1915* (New York: Oxford University Press, 1999); and Nancy L. Gallagher, *Breeding Better Vermonters: The Eugenics Movement in the Green Mountain State* (University Press of New England, 1999).

11. Kevles, *In the Name of Eugenics*, p. 62.

12. *New York Times*, 30 November 1993.

13. E. Richard Gold, *Body Parts: Property Rights and the Ownership of Human Biological Materials* (Washington, DC: Georgetown University Press, 1996), p. 148.

That is, commodification entails that all modes of valuation are commensurate with economic valuation.

Margaret Jane Radin, Stanford University professor of law, has done seminal work on commodification theory. Both in her work *Reinterpreting Property*[14] and her more recent volume *Contested Commodities: The Trouble with Trade in Sex, Children, Body Parts, and Other Things*,[15] Radin examines what she calls "commodification as a worldview."[16]

> In universal commodification, the value of a commodity (from the social point of view) is defined as its exchange value, often referred to as market value, when it is traded in a laissez-faire market — or hypothetically traded in a hypothetical laissez-faire market. Valuation in terms of dollars implies that all commodities are fungible and commensurable — capable of being reduced to money without changing in value, and completely interchangeable with every other commodity in terms of exchange value.[17]

Radin denies that all values or modes of valuation are commensurable with market values. In fact, the burden of her work is to demonstrate that some forms of commodification are clearly wrong. For instance, a market in babies is immoral in Radin's view.[18] When a baby is bought and sold in the marketplace — i.e., becomes a commodity — that child's personal traits or attributes (sex, eye color, I.Q., predicted height, etc.) also become commodified.[19] Furthermore, according to Radin, commodification of the infant is *ipso facto* a form of commodification of the future person (e.g., the academic, the homemaker, the career woman, etc.) the baby will become. Baby giving (adoption) is nonproblematic for Radin. In fact, giving up a child for adoption may be seen as a commendable form of altruism in those cases where the adoption results in good to the child and contributes to the adoptive parents' lives.

In 1980, the United States Supreme Court upheld the patenting of organic life. In 1986, the U.S. Patent and Trademark Office issued 37 patents on

---

14. Margaret Jane Radin, *Reinterpreting Property* (Chicago: The University of Chicago Press, 1993).

15. Margaret Jane Radin, *Contested Commodities: The Trouble with Trade in Sex, Children, Body Parts, and Other Things* (Cambridge, MA: Harvard University Press, 1996).

16. Radin, *Contested Commodities*, pp. 1-14.

17. Radin, *Contested Commodities*, p. 3.

18. For the purposes of her argument, Radin distinguishes between surrogacy and a woman selling her own baby, though it is clear that Radin finds paid surrogacy unethical. See Radin, *Contested Commodities*, pp. 142ff.

19. Radin, *Contested Commodities*, p. 137.

genes. Incyte Pharmaceutical holds about 500 gene patents, the largest private holdings in the nation. Celera Genomics, the firm that helped map the genome, recently applied for patents on 7,000 SNPS (bits of genes). Since patents are limited monopoly rights to control the sales, use, and manufacture of the genes or any products made from those genes, they entail commodification, pure and simple. Economic values are not commensurate with the way we ought to value human genes. Genes are God's donation to us all. They are not to be the objects of private, commercial biotechnology.

Radin points out that by 1993 there were nearly 1,300 biotechnology firms in the United States, employing over 80,000 individuals, with sales of $6 billion annually.[20] The Biotechnology Industry Organization (BIO), whose purpose it is to lobby legislators on issues related to the industry, presently maintains an office in Washington, D.C. BIO represents over six hundred biotechnology companies, academic institutions, state biotechnology centers, and other organizations in forty-seven states and more than twenty countries.[21]

In the United Kingdom there are over five hundred public and private biotech organizations. Oliver adds that "others with a sizable per capita biotech industry include Austria, Belgium, Denmark, Finland, France, Germany, Ireland, Italy, The Netherlands, Norway, Poland, Spain, Sweden, and Switzerland. Belgium has some of Europe's most entrepreneurial biotechs, while Denmark has been a leader in developing biotech regulations."[22]

In sum, there are over two thousand biotechnology organizations in the United States and more than a thousand in the European Union. Investment in U.S. biotechnology increased twelve percent in 1997, from $83 billion to $93 billion. Biotechnology is one of the fastest-growing business segments around the globe.[23]

Biotechnology may be, as has been suggested, a set of techniques; but at the outset of the third millennium, we must also acknowledge that biotechnology is a recognizable, market-driven, capitalism-wedded, corpus of businesses whose primary purpose is to make profits.[24] As Colin Ratledge stated,

20. Radin, *Contested Commodities,* p. 1.
21. The Biotechnology Industry Organization (BIO) is the world's largest trade organization to serve and represent the emerging global biotechnology industry. According to Carl Feldbaum, president of BIO, biotechnology has become a significant economic force, with more than 1,300 companies, nearly $13 billion in annual revenues and more than 100,000 people on its direct payroll. *BIO Editors' and Reporters' Guide to Biotechnology 1996-1997* (Washington, DC: Biotechnology Industry Organization, 1996), p. 1.
22. Oliver, *The Coming Biotech Age,* p. 208.
23. Oliver, *The Coming Biotech Age,* p. 36.
24. It should be said that "making money" is not necessarily negative. In fact, the

"Biotechnology is not then a science: it is a means of applying science for the benefit of man and society. In practice, this means that biotechnology is used to make money — or in certain instances — to save money."[25] Biotechnology promises to remain quite lucrative. *Genetic Engineering News* noted in December 1995 that U.S. sales of biotechnology products are expected to grow at an average annual rate of approximately twelve percent, from $10 billion in 1996 to $32 billion in 2006.[26]

The narrative of biotechnology is varied and complex. Nevertheless, after tracing the lineaments of its history, several preliminary observations seem warranted. First, the evolution of biotechnology is part of a larger historico-cultural revolution. Second, the "machine" metaphor became a powerful conceptual apparatus for the biotechnological revolution. Third, a shift in the mode of valuing living things occurred during the ascendency of biotechnology. The engineering model applied to biology contained an implicit paradigm shift in the way organisms, tissues, cells, and genes were valued. In short, these and other intellectual developments of modernity (such as naturalism) set the stage for the pervasive philosophical materialism of our own era. Philosophical materialism combined with capitalistic fervor results in commodification of the human body and its parts.

---

most charitable reading of the biotechnology industry's goals leads one to the conclusion that the industry has very altruistic goals. For instance, Amgen (Applied Molecular Genetics), the world's largest independent biotechnology company, identifies as its mission "to be the world leader in developing and delivering important, cost-effective therapeutics based on advances in cellular and molecular biology." Similarly, Amgen declares, "Our primary purpose is to bring meaningful improvement to the lives of patients through our products." "Amgen's Mission, Goals & Values," http://wwwext.Amgen.com/cgi-bin/genobject/amgenValues/tigo_8AIFId.

25. Colin Ratledge, "Biotechnology: The Socio-economic Revolution? A Synoptic View of the World Status of Biotechnology," in *Biotechnology: Economic and Social Aspects,* ed. E. J. Da Silva, C. Ratledge, and A. Sasson (Cambridge: Cambridge University Press, 1992), p. 1. Interestingly, Ratledge maintains that "altruistic biotechnology does not exist or if it does it simply consumes money and does not generate it" (p. 3).

26. "Double-Digit Growth Predicted for Biotechnology Products in the Next Decade," *Genetic Engineering News* (December 1996): 6. The entrepreneurial aspects of biotechnology most assuredly affect the methodology used by those who function within its environs. For instance, Kornberg and others point out that the so-called "targeted research" agenda may have deleterious consequences on future discovery. Cf. "Pros and Cons of Biotech Ventures," ch. 8 of Arthur Kornberg, *The Golden Helix* (Sausalito, CA: University Science Books, 1995), pp. 231-58.

What Is Complicity?

Another important ethical challenge we face is defining and responding to the problem of complicity. The debate over stem cell research has embodied a very anemic view of complicity. The National Institutes of Health essentially erected a firewall between the destruction of an embryo and research on human embryonic stem cells. According to NIH, tax funds may not be used to pay for the act of destroying a human embryo for research purposes. Presumably, this act is morally sullied enough to warrant restricting public funds for such purposes. Tax funds may, however, be used to fund research following the isolation of the stem cell. By separating the act of killing the embryo from the research using the embryo's stem cells, NIH argued that researchers were not morally complicit in the act of killing the embryo. This is morally unacceptable smoke and mirrors.

The assisted suicide debate has encouraged an equally anemic notion of complicity. If, under the consumer model of medicine, the patient wants assistance in ending his or her life, the physician is bound to follow the infallible customer ("the customer is always right"). This argument suggests that, in those cases where a competent patient requests assistance in dying, the physician is somehow not to be viewed as an accomplice.

Are persons who are removed from an act either by financial gain or distance rendered morally sanitized from abhorrent acts of which they are well aware? Two examples should suffice to answer this question. First, consider persons who may be removed from a good act but believe they are accomplices in that act. Take, for instance, makers of the ceramic tiles which form the skin of the space shuttles. Most workers who produce those tiles do not work for NASA and have no organic ties to the U.S. government. They work for private companies who sell their products to the government. Their own work is not tax-funded, though the work of NASA is. When a successful shuttle expedition is celebrated, those same workers celebrate the success as, in part, their own success. That is, they believe they are accomplices in the success of the space shuttle. It would not be unusual to find them wearing t-shirts saying something like, "I made tiles for the space shuttle." Even though they are not employees of NASA, they feel that they are complicit in the nation's aeronautic success. If complicity works in that direction, why would it not work in the other direction? If individuals are complicit in good acts, even though they are removed from those acts through funding mechanisms and distance, are they not also complicit in bad acts of a relevantly similar nature? How, therefore, can NIH hope to separate the morality of destroying the embryo from the research using the embryo's stem cells?

190

The second example comes from the arena of research ethics. The World Medical Association's Declaration of Helsinki II, adopted at the 29th World Medical Assembly held in Tokyo, Japan, October 1975, states: "In publication of the results of his or her research, the doctor is obliged to preserve the accuracy of the results. *Reports of experimentation not in accordance with the principles laid down in this Declaration should not be accepted for publication*" (Section I, Article 8, italics added). The rationale for prohibiting publication of research which violates the Declaration and, therefore, the rights of human research subjects, is that such research, no matter how scientifically valuable, would be morally tainted. That is another way of saying that those who publish findings from unethical research are morally complicit in that research. So, for instance, H. A. Israel and W. E. Seidelman reject the use of the *Pernkopf Atlas* text in medical education, but not because it is a substandard text. In fact, it is, by some accounts, a highly superior anatomy textbook. The problem with the text is that it was written by a Nazi doctor who committed moral atrocities against his patients in order to research the book.[27] Pernkopf and all who would use that text would to some, even if limited, degree be complicit in the barbarous acts which provided some of the data for the anatomy text.

Responding to the Challenges

There are other questions we face in this new biotech century. Regulating biotechnology will require careful attention. On the one hand, the promise of biotechnology will be realized through promiscuous research and development. On the other hand, biotechnology holds such great potentials for evil (e.g., bioweaponry and genocide), not to mention the problem of unintended deleterious consequences (e.g., environmental contamination), that some form of regulation must be in place if tragedies are to be avoided. The difficulties in crafting appropriate regulations and in establishing effective regulatory oversight are many. Not only must oversight include both publicly-funded and privately-funded research, but oversight must include a world-wide research community within a global marketplace. The task is truly daunting and the stakes are extraordinarily high.

Thus, if we are to face the challenges before us, nothing less than an intentionally collaborative approach will do. First, we must find common cause

---

27. H. A. Israel and W. E. Seidelman, "Nazi Origins of an Anatomy Text: *The Pernkopf Atlas,*" *Journal of the American Medical Association* 276 (1996): 1633-34. My thanks to William P. Cheshire, M.D., for bringing this article to my attention.

and build consensus around certain tried and tested standards. For many of us, that will mean arriving at theological conclusions on those issues which are derived from a historical-grammatical exegesis of the scriptures of the Old and New Testaments. Our confession is that God has not left us without sufficient revelation which may be applied to matters such as those before us. At the same time, we realize that we live in a pluralistic culture in which not everyone shares our confidence in and commitment to the Bible. Those who are committed to the biblical witness may on many points substantially agree with those who are committed to monotheism. Christians, Jews, and Muslims can find common ground on which to address a host of today's questions. Because they are committed to a similar understanding of human responsibility before God and share similar understandings of the nature of the human person before God, they should be able to address the issues enumerated above with a common voice.

The consensus might be even broader and more comprehensive than that — for example, within a Hippocratic morality. The covenantal quality of the Hippocratic Oath, with its exposition of physicians' duties to patients and to profession, has been the foundation of medicine from antiquity. In fact, one might well argue that medicine divorced from Hippocratism ceases to be medicine. It might be a technique, a form of "body plumbing," but it is not medicine unless it is Hippocratic. We must revive Hippocratic medicine if we are to see medicine serve humanity in this century. Raw technique tends to be self-serving and self-propagating. Hippocratism is a way of understanding the physician's obligations and the patient's good as the heart of the clinical encounter. Hippocratic medicine is more than technique, it is a covenant relationship with transcendent values which govern it. As Nigel Cameron has well said, "Christians and others who share the Hippocratic values are called to follow in the footsteps of the first Hippocratic physicians, and to assume the dissident and reforming role which once was theirs."[28] This task may in fact be the greatest challenge ahead.

In summary, the following are recommendations for prophetic engagement in the biotech century:

- We must engage in multidisciplinary collaboration across the widest possible spectrum. Philosophers, theologians, physicians, lawyers, biblical scholars, priests, rabbis, and imams should bring their various skills and traditions to bear on the thorny questions before us.

28. Nigel M. de S. Cameron, *The New Medicine: The Revolution in Technology and Ethics* (London: Hodder & Stoughton, 1991), p. 167.

- In order to faithfully fulfill their prophetic role, churches will have to make it a priority to teach Christian ethics in general and bioethics in particular. Synagogues and mosques must give similar attention to bioethics.
- Pastors should preach and teach biblical anthropology since all of these biotechnologies impact human beings positively or negatively.
- Seminaries must carve out either curricular or extracurricular opportunities for students to learn about the developments in biotechnology and be provided skills to interpret those technologies from a Christian worldview perspective.
- Church educators must reprioritize the educational ministry of the church, giving increased attention to bioethical issues, including biotechnological issues.
- Evangelicals should increase funding and personnel resources to agencies and commissions that have a direct impact on biotechnology policy, including international policy work.
- Christian students should be encouraged to pursue vocations in biotechnology and the sciences. Individuals can impact biotechnology at the local level by bringing their convictions to bear in their own vocations.

We must advance, not retreat, in the face of today's challenges.

Index